PERGAMON INTERNATIONAL LIBRARY
of Science, Technology, Engineering and Social Studies
The 1000-volume original paperback library in aid of education, industrial training and the enjoyment of leisure
Publisher: Robert Maxwell, M.C.

The Historical Supernovae

Other Titles of Interest:

BEER:	Vistas in Astronomy (review journal)
BOWLER:	Gravitation and Relativity
ELGAROY:	Solar Noise Storms
GLASBY:	The Nebular Variables
HEY:	The Radio Universe , 2nd edition
HILLAS:	Cosmic Rays
MEADOWS:	Early Solar Physics
PACHOLCZYK:	Radio Galaxies
REDDISH:	Stellar Formation
ROWAN-ROBINSON:	Far Infrared Astronomy

The Historical Supernovae

BY

DAVID H. CLARK
Mullard Space Science Laboratory
University College London

AND

F. RICHARD STEPHENSON
Institute of Lunar and Planetary Sciences
University of Newcastle upon Tyne

PERGAMON PRESS
OXFORD · NEW YORK · TORONTO · SYDNEY · PARIS · FRANKFURT

U.K.	Pergamon Press Ltd., Headington Hill Hall, Oxford OX3 0BW, England
U.S.A.	Pergamon Press Inc., Maxwell House, Fairview Park, Elmsford, New York 10523, U.S.A.
CANADA	Pergamon of Canada, Suite 104, 150 Consumers Road, Willowdale, Ontario M2 J1P9, Canada
AUSTRALIA	Pergamon Press (Aust.) Pty. Ltd., P.O. Box 544, Potts Point, N.S.W. 2011, Australia
FRANCE	Pergamon Press SARL, 24 rue des Ecoles, 75240 Paris, Cedex 05, France
FEDERAL REPUBLIC OF GERMANY	Pergamon Press GmbH, 6242 Kronberg-Taunus, Pferdstrasse 1, Federal Republic of Germany

---

First edition 1977

Reprinted 1979

Library of Congress Cataloging in Publication Data

Clark, David H
The historical supernovae.

1. Supernovae. 2. Stars--Observations--History.
I. Stephenson, Francis Richard, Joint author. II. Title.
QB841.C58 1977 532.8'446 76-44364
ISBN 0-08-020914-9 (Hardcover)
0-08-021639-0 (Flexicover)

In order to make this volume available as economically and rapidly as possible the author's typescript has been reproduced in its original form. This method unfortunately has its typographical limitations but it is hoped that they in no way distract the reader.

Printed in Great Britain by A. Wheaton & Co. Ltd, Exeter

To Suzanne and Ellen

"The investigation of the remnants of supernovae and their relation to historical records, both written and unwritten, will be one of the most fascinating tasks awaiting the next generation of astronomers ......"

Fritz Zwicky - joint instigator of the term supernova, and a pioneer of modern supernova surveys;
(in "Stellar Structure", Volume VIII of "Stars and Stellar Systems", Ed. Aller and McLauglin. University of Chicago Press. 1965)

"The extent to which the Chinese records of guest stars remain of living interest to current astronomical research may be seen in the field of radio-astronomy, where during the past few years great additions to knowledge have been made......The rapid upsurge of this new and powerful method of study of the birth and death of stars....makes urgently necessary the reduction of the information contained in the ancient and medieval Chinese texts to a form utilisable by modern astronomers in all lands. For this purpose, however, collaboration between competent sinologists and practical astronomers and radio astronomers is indispensable."

Joseph Needham, F.R.S. - distinguished historian of Chinese Science;
(in "Mathematics and the Sciences of the Heavens and the Earth", Volume 3 of "Science and Civilisation in China". Cambridge University Press. 1959)

CONTENTS

PREFACE

In writing this book it has been our intention to approach the subject of supernovae by way of the historical records, with particular reference to the history of the Far East. The book therefore is an interdisciplinary study suitable for students of Astronomy, History of Science or Sinology, as well as the non-specialist.

When directing a work to such a wide audience, clearly allowance must be made for the varying backgrounds and interests of readers. Consequently Chapters 1 to 4 contain much introductory material to assist in the understanding of the remaining eight chapters.

The impact of the study of historical supernovae on modern astrophysics has been very significant. Readers need only be reminded that many of the most dramatic discoveries in Astronomy during the past few years, such as pulsars, black holes, X-ray binaries, etc., are directly linked with supernova explosions. We have not hesitated to describe the historical records of supernovae as "among the most valuable legacies which the ancient world has bequeathed to modern science".

During the preparation of this book we have benefited much from collaboration with Mr. A. C. Barnes, School of Oriental Studies, University of Durham, and Dr. J. L. Caswell, Radiophysics Division, CSIRO, Australia. It is a pleasure to acknowledge our gratitude to both colleagues. We are indebted to Professor A. J. Meadows, Department of Astronomy and History of Science, University of Leicester, for reading the manuscript and offering several valuable criticisms and suggestions, and also to Professor W. H. McCrea, Astronomy Centre, University of Sussex, for collaboration on some of the material in Chapter 12.

Our special thanks are due to Miss Jane Salton for typing the manuscript. The help of Dr. T. J. Saunders and Dr. A. J. Woodman, both of the Department of Classics, University of Newcastle upon Tyne, with difficult Latin texts is also gratefully acknowledged.

Finally we wish to thank our families for their continued encouragement during the preparation of this book.

David H. Clark

F. Richard Stephenson

November, 1976.

ACKNOWLEDGEMENTS

We are grateful to the following for permission to publish copyright material -

The British Museum, London - Plate 1.
The Royal Scottish Museum, Edinburgh - Plate 5.
University of Chicago Press - Plates 6, 7 and 9 are from "An Optical Atlas of Galactic Supernova Remnants" by van den Bergh, Marscher, and Terzian, The Astrophysical Journal Supplement No. 227.

Chapter 1

NEW STARS — NOVAE AND SUPERNOVAE

The vast majority of stars in a typical galaxy such as our own (the Milky Way) are extremely stable, and produce a remarkably steady output of radiation over millions of years. In marked contrast are the Novae and Supernovae - stars which spontaneously explode with a spectacular and rapid increase in brightness, so that at maximum they rival the brightest stars in the heavens, before gradually fading into insignificance. At its height a nova may be intrinsically a million times as bright as the Sun, with a supernova at its height being several thousand times as bright again and emitting as much energy as all the other stars in the galaxy combined. The term "nova" means literally a new star, and was introduced for such objects before the realisation that a star actually existed at the site of the nova or supernova prior to the outburst.

A supernova is now recognised as the violent end to the evolution of a certain type of star. Although the exact reason why some stars undergo such catastrophic self-destruction is not fully understood, the evolutionary path leading up to at least one type of supernova event may be described quite simply. As a nascent star forms from a cloud of gas and dust collapsing under the action of gravitational forces, the temperature in the interior of the cloud increases to the point ($\approx 10^7$ °K) where certain nuclear reactions can take place. At first the dominant nuclear reaction is the so-called proton-proton chain reaction, the net effect of which is the fusion of hydrogen to form helium with the release of energy. The liberation of this thermonuclear energy halts the gravitational contraction, and the star settles down to the relatively stable state in which it spends most of its active life - "burning" hydrogen in the central core to balance the energy radiated from the surface.

Although the nuclear fuel reserves of a star are enormous, they are not unlimited. When the core hydrogen is expended, a renewed contraction of the core raises the temperature to the level ($\approx 2 \times 10^8$ °K) required to initiate the burning of helium to form carbon and oxygen, accompanied by hydrogen burning in the surrounding shell. Later stages follow involving the fusion of successively heavier elements all the way to iron, when no further reactions can extract energy. Since the energy yield per fusion and the abundance of the reacting elements both decrease at each successive stage of nuclear burning, the star squanders its remaining fuel reserves at an ever increasing rate as it approaches its final destruction. Several stages of nuclear burning occur simultaneously in the same star, with the core being most advanced and reaching the iron end-point while the outer portion of the stellar envelope is still burning hydrogen, with intervening zones at intermediate stages.

What follows when no further nuclear energy is available in the stellar core depends on the mass of the star. Those stars with core masses less than about 1.4 $M_{\odot}$ ($M_{\odot} \equiv$ the mass of our Sun) undergo a relatively controlled collapse to a state of very high density ($\simeq 10^6$ grams per cm^3), forming a so-called white dwarf. By contrast, stars with cores heavier than 1.4 $M_{\odot}$ experience rapid collapse accompanied by the release of gravitational energy. For particularly massive stars the core mass and its kinetic energy of implosion may drive the system beyond nuclear densities to complete gravitational collapse to a black hole. However the collapse of the cores of less massive imploding stars (those with masses in the range 1.4 to 3.5 $M_{\odot}$) will be halted at nuclear density to give a rapidly rotating neutron star (density $\simeq 10^{14}$ grams per cm^3), which may be observable as a pulsar emitting pulses of radiation over a wide range of frequencies at intervals of a fraction of a second. The sudden stopping of the collapsing core at nuclear density yields an enormous temperature ($\simeq 10^{12}$ ^{o}K) and high pressure, producing an outward moving shock wave. The stellar envelope, still containing elements lighter than iron from intermediate stages of nuclear burning, and collapsing behind the core, will suddenly experience this shock with resulting rapid compression and sharp increase in temperature. The nuclear reaction involving the fusion of oxygen-16 nuclei to form silicon (at a temperature of about 10^9 ^{o}K) is potentially explosive, and the initiation of this reaction in the collapsing envelope is believed to contribute to the explosive ejection of a significant fraction of the stellar mass. These final stages of the evolution of a massive star are depicted schematically in Fig. 1.1.

A supernova event thus results in the sudden implosion of the core (on a time-scale of seconds) to leave a rapidly rotating collapsed stellar remnant, the explosive ejection of the stellar envelope broken up into filamentary structure flying off with velocities of thousands of kilometres per second, and the release of vast quantities of energy of the order of 10^{51} ergs (equivalent to about 10^{28} 1-Megaton hydrogen bomb explosions).

Supernovae are of tremendous interest to astrophysicists, not only because they represent the most spectacular of stellar events, but because the remnants and ejecta of such explosions are amongst the most unusual and exciting of astrophysical objects and phenomena. In addition to pulsars and black holes, supernova explosions are believed to be responsible for the production of high-velocity "runaway" stars hurtling through the galaxy at speeds approaching a million miles per hour, the high-energy cosmic rays continually bombarding our planet, the heavy elements, spectacular expanding nebulosities which are amongst the most beautiful objects in the heavens, extended sources of radio emission, possibly gravitational radiation, and probably most of the galactic X-ray sources.

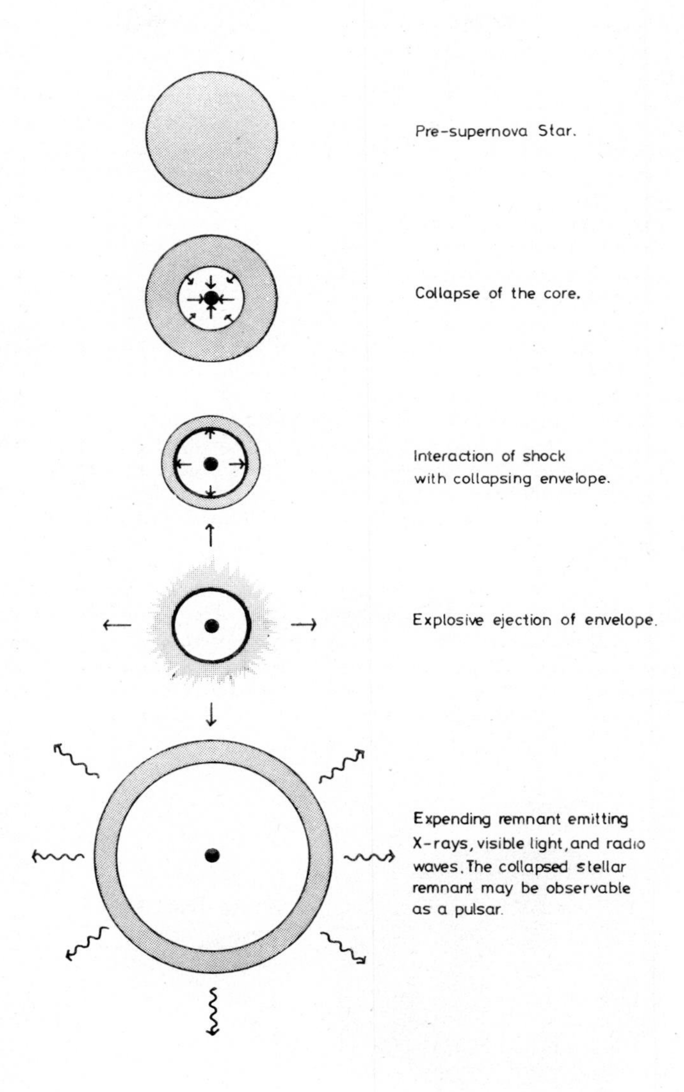

Fig. 1.1. A schematic representation of the self-destruction of a star in a supernova event.

By contrast with a supernova, which is a "one-off" event heralding the final destruction of a star, a nova is believed to be merely a temporary departure of certain stars from the well-established stellar evolutionary path, and may in fact be a repetitive phenomenon with the interval between outbursts being up to several thousand years. In each nova explosion only a very small fraction of the star's mass (possibly no more than one hundredth of one percent) is believed to be ejected, and some of this may eventually be drawn back onto the star. Although the exact reason why certain stars experience nova explosions remains open to debate, the most widely held theory is based on the now convincing observational evidence that a prenova is a close binary system. A close binary system is a pair of stars which are close enough for mass transfer between them to occur at certain times during the normal course of their evolution. Any such system containing a white dwarf and a normal companion star is a potential nova. During an expansion phase of the companion star, material (mainly hydrogen) is transferred to the atmosphere of the white dwarf, where under the action of the strong gravitational field it is compressed and heated to the ignition temperature for hydrogen burning reactions. Thereafter thermonuclear runaway could produce the explosive ejection of matter and the radiation of energy seen as the nova event. The interchange of material within a close binary system preceding a nova outburst is depicted in Fig. 1.2.

A nova explosion resembles a supernova in many respects, but is on a much smaller scale. A typical nova lasts days or at most weeks rather than months, and the energy released is less than one ten-thousandth of that in a supernova explosion. The expanding shell of gas from the explosion dissipates in a few decades, leaving the binary system showing little change from its original

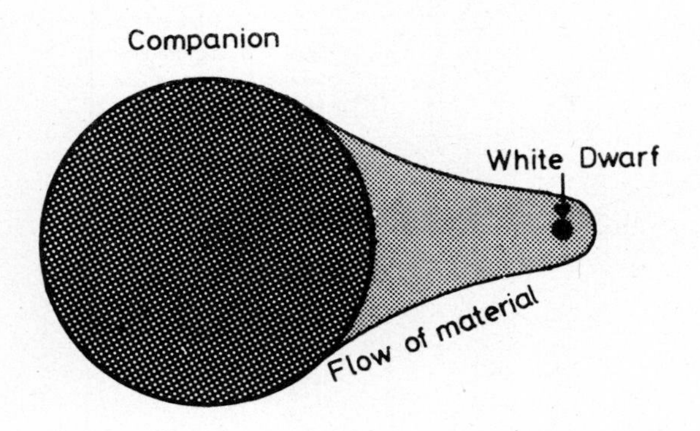

Fig. 1.2. The interchange of material within a close binary system preceding a nova outburst.

condition. Novae are much more frequent than supernovae, and for a short time a nearby nova can shine among the brightest stars in the sky. There have been six bright novae in the present century, appearing in 1901, 1918, 1925, 1934, 1942 and 1975. All temporarily rivalled first magnitude stars, and for a few days the 1918 nova was almost as bright as Sirius (the brightest permanent star in the sky).

Although novae and supernovae differ in their origin, in the scale of outburst, and indeed in their frequency of occurrence, the existence of supernovae as events distinct from common novae has been recognised only in the last forty years.

There have been many outbursts of an apparently stellar nature recorded throughout history. Most early observations of "new stars" were made in the Far East - China, Japan and Korea. In ancient and medieval Europe and the Arab Lands there seems to have been little interest in such phenomena, partly due to the widespread influence of the Aristotelian doctrine of a perfect changeless celestial sphere and partly the result of sheer inability to recognise a new star-like object (see Chapter 2). This situation was not to change until the 16th century.

In the Far East, professional astronomers/astrologers were employed by the ruler to maintain a constant watch of the sky, and to report and interpret any unusual events which might happen. Many such events are reported in the astronomical treatises of the various dynastic histories. We have remarkably detailed records from China going back to about 200 BC, but unfortunately, probably as the result of the famous "Burning of the Books" at the direction of Ch'in Shih-huang, the first true emperor of China, in 213 BC, there is very little before that date. In Japan and Korea, regular astronomical observations began around AD 1000, so that after this date we find frequent duplication. The historical background of the Far Eastern observations will be further discussed in Chapter 2.

Ideally, what we would like to find among the historical records are accounts of new stars describing in detail the position, changes in brightness, colour and duration of visibility. Such completeness is seldom realised in practice, but it is important to take full advantage of whatever information we can obtain from the early observations. Several of the historical records of new stars make reference to extreme brightness. However it is not possible to make a distinction between novae and supernovae in our own galaxy on this basis alone, since the apparent luminosity of novae occurring close to the Earth would not differ from that of the intrinsically very much brighter supernovae at greater distance.

There were two principal reasons for originally suggesting that there existed stellar explosions thousands of times as bright as common novae. The first of these was based on the observations made of new stars in extragalactic stellar systems. In 1885 a temporary star was observed near the nucleus of Messier 31 which nearly equalled the apparent brightness of that galaxy, and since that time several hundred new stars have been detected in external galaxies. If the number of such events in different apparent brightness ranges is plotted for a few nearby galaxies, it immediately becomes apparent that there is a small, but distinct, group of extreme brightness (see Fig. 1.3). (The scale of stellar magnitudes is a logarithmic one in which negative numbers indicate a greater brightness than positive numbers, and in which each magnitude is some 2.5 times brighter than the one below it. The observed magnitude of a star is called its apparent magnitude. For example, zero apparent magnitude is the brightness of Vega, and -4 apparent magnitude is the brightness of Venus. To compare one star with another we need to know their intrinsic magnitudes, such as would be obtained if the stars were placed at a standard distance. The distance chosen is 32.6 light years, = 10pc, and the magnitude at this distance is termed the absolute magnitude.)

The second reason for suggesting the existence of supernovae was related to the new star of AD 1572, described in some detail and accurately positioned by the eminent Danish astronomer Tycho Brahe. Ordinary novae show an increase from pre-nova to maximum brightness typically of about 15 magnitudes. The new star of 1572 was for a few days as bright as Venus, about apparent magnitude -4. If an ordinary nova, it must have been within a few tens of light years of the Earth, and its stellar remnant should now not be fainter than +11. However no such object could be found, and it was concluded that the new star of AD 1572 must have been at large distance and with extreme intrinsic brightness at maximum.

Thus, in 1937, the term supernova was proposed for the distinct group of new stars of extreme intrinsic brightness by Zwicky and Baade, two of the pioneers of supernova research.

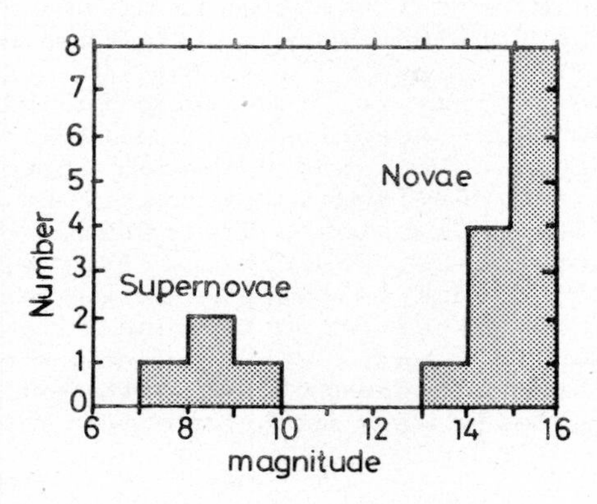

Fig. 1.3. A distribution of bright novae and supernovae in nearby galaxies (from Zwicky 1965).

Three other galactic new stars recorded historically have for some time been recognised to be supernovae. A new star which appeared in the year AD 1604 in the constellation of Ophiuchus was carefully observed for about 12 months by many of the leading astronomers of Europe and independently by the court astronomers of Korea and China. The celestial location of the star was so accurately measured by Kepler and Fabricius (to better than a minute of arc) that Baade (1943) had no difficulty in locating a small patch of nebulosity which clearly represented its optical remnant.

The new star of AD 1054 was recorded only by the Chinese and Japanese. Combining the various reports, we learn that the star was visible in daylight for a total of 23 days, with a total period of visibility exceeding 22 months. That this event was associated with the Crab Nebula was possibly first suggested by Lundmark (1921) ; certainly the rate of expansion of the optical filaments of the nebula gives a date of outburst in close agreement with AD 1054, and there is every reason to believe that the Crab Nebula is one of the few select supernova remnants of precisely known age.

The new star which appeared in AD 1006 is the only such star known to be recorded in European and Arabic history before the Renaissance. It was also extensively observed by the Chinese and Japanese astronomers. Many texts comment on the extreme brilliance of the star. For example ; "It shone so brightly that objects could be seen by its light" (China) ; "Its rays on the Earth were like the rays of the Moon" (Iraq) ; "Its light illuminated the horizon and ... its brightness was a little more than a quarter of the brightness of the Moon" (Iraq). One Chinese record states that the star was followed for several years. Goldstein (1965) not unrealistically estimated the apparent magnitude at maximum as between -8 and -10.

There can be little doubt that the long-duration new stars of extreme brilliance sighted in AD 1006, AD 1054, AD 1572 and AD 1604 were supernovae. (They will be discussed in detail in later chapters.) At least four supernovae occurring on the near-side of the galaxy during the past millennium would seem to place a lower limit on the frequency of galactic supernovae of about one per 120 years. However it must be stressed that the short interval between some actual sightings of supernovae can give a grossly misleading impression of the frequency of events because of the extreme distances involved. For example, the AD 1006 supernova occurred at a distance of about 3 thousand light years from the Earth, while that of AD 1054 occurred at a distance of over 6 thousand light years, so that although first sighted on Earth within half a century of each other the outbursts were actually separated in time by over 3000 years. Despite this cautionary note, there is growing evidence that in galaxies such as our own supernovae occur, on average, about once per century.

The fact that few supernovae have been recorded historically is

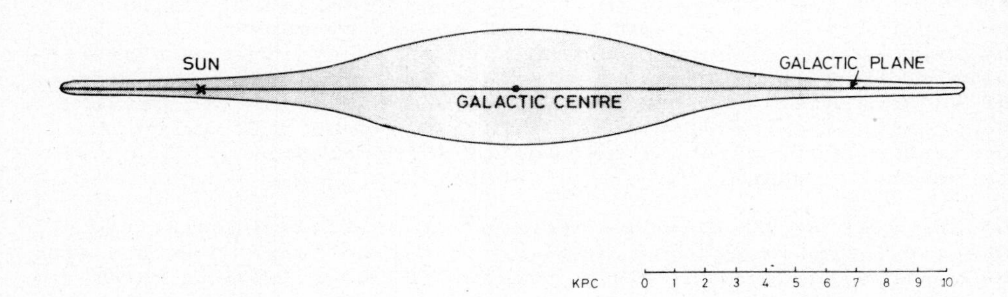

Fig. 1.4. A schematic edge-on view of the Galaxy.

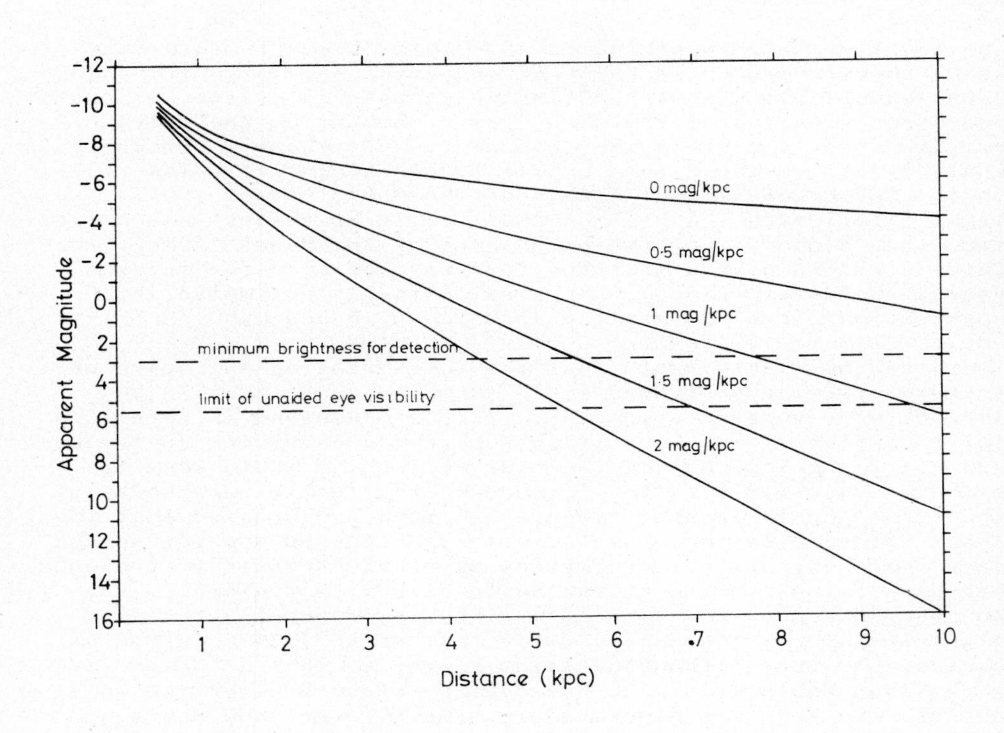

Fig. 1.5. Curves of apparent (or observed) magnitude of a supernova (absolute magnitude -19), as a function of distance for selected values of interstellar absorption between 0 and 2 magnitudes per kpc.

partly a consequence of the high degree of optical obscuration produced by dust clouds which permeate the whole galaxy and blanket out many such events. A large portion of the Milky Way lying towards the galactic centre (which is located in the constellation of Sagittarius) is effectively screened from us optically, The situation is not improved by the Sun's location almost exactly on the galactic plane where the concentration of dust is highest - see Fig. 1.4. (The distance scale in Fig. 1.4 is in kiloparsecs, kpc, where one parsec is the distance of a star with an angle subtended by the radius of the Earth's orbit of one second of arc. One parsec is equivalent to 3.263 light-years.) In spite of its extreme intrinsic brilliance, a supernova in our galaxy is unlikely to be detected unless it lies at a considerable distance to the north or south of the galactic plane or else happens to occur in the Sun's neighbourhood. This is highlighted in Fig. 1.5, which plots the diminishing apparent magnitude for a supernova of maximum absolute magnitude -19, with increasing distance from the Earth, and for selected values of apparent interstellar absorption.

In contrast, external galaxies are visible to us from all directions, and because of their small angular size can be studied as a unit. These are just two of the reasons why the extra-galactic supernova searches at Mt. Palomar, Asiago (Italy), and elsewhere have been so successful. As many as three or even four supernovae have been detected in each of a number of external galaxies in the past few decades, whereas none has been seen in our own galaxy since AD 1604, 5 years before Galileo Galilei first applied the telescope to astronomical research.

The extragalactic surveys have provided important new insights into the nature of supernovae. In particular, it has been found that two main types of supernovae can be recognised on the basis of absolute luminosities, light curves, and spectra. The so-called Type I supernovae typically have an absolute magnitude at maximum of -19, and after an initial drop of about 3 magnitudes in 20 to 30 days the light curve shows an approximately exponential decay (see Fig. 1.6.) Type II supernovae are rather more individualistic, although a common feature of their light curves is a "shoulder" after maximum, followed by a rather rapid decline (see Fig. 1.7.) The composite supernova light curves of Figs. 1.6 and 1.7 may be compared with those for common novae shown in Fig. 1.8. Type I supernovae are believed to be produced by stars of mass slightly greater than one solar mass, while Type II supernovae are the progeny of very much more massive stars and may involve the ejection of several solar masses in the explosion and there are clear indications that ejections at different velocities occur.

The maximum apparent magnitude for a supernova may be crudely estimated from an historical record giving only the period of visibility. As noted earlier, and depicted in Fig. 1.6, a Type I supernova light-curve falls about 3 magnitudes in 20 to 30 days; the subsequent decline from maximum absolute magnitude is approximately 4 magnitudes after 100 days and approximately

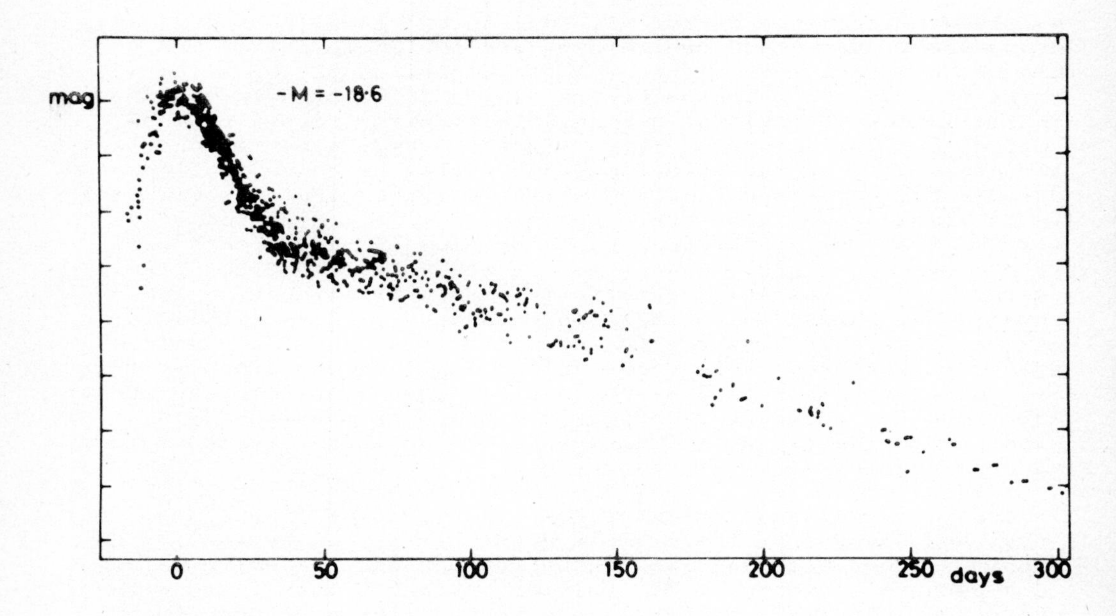

Fig. 1.6. Composite blue light curve obtained by the fitting of the observations of 38 Type I supernovae. One magnitude intervals are marked on the ordinate. (From Barbon et al, 1974a).

7 magnitudes after 300 days, etc. The limit of naked-eye visibility is about magnitude +5.5. Thus, for example, a recorded duration of visibility of 300 days would imply a maximum absolute magnitude of -1.5 assuming typical Type I supernova behaviour. The technique is of course highly subjective, and cannot be used assuming Type II parameters because of the uncertain behaviour of Type II supernovae beyond about 100 days. Nevertheless, we will have cause to apply the technique later to give preliminary estimates for the brightness and thus the distance of certain supernovae.

Lacking any telescopic observations of galactic supernovae, it is not surprising that today astrophysicists are concentrating much effort on the study of supernova remnants (SNRs). A supernova may produce a number of possible observable remnants ; the main ones being a pulsar, an expanding optical nebulosity, an extended radio source, and an extended X-ray source (see Chapter 4). The shock wave expanding from the initial outburst is usually found to delineate a region of radio emission, while the heated interstellar material swept up by the shock wave may act as an observable source of X-rays. Obscuration means that only the nearby supernova remnants can be observed at optical and

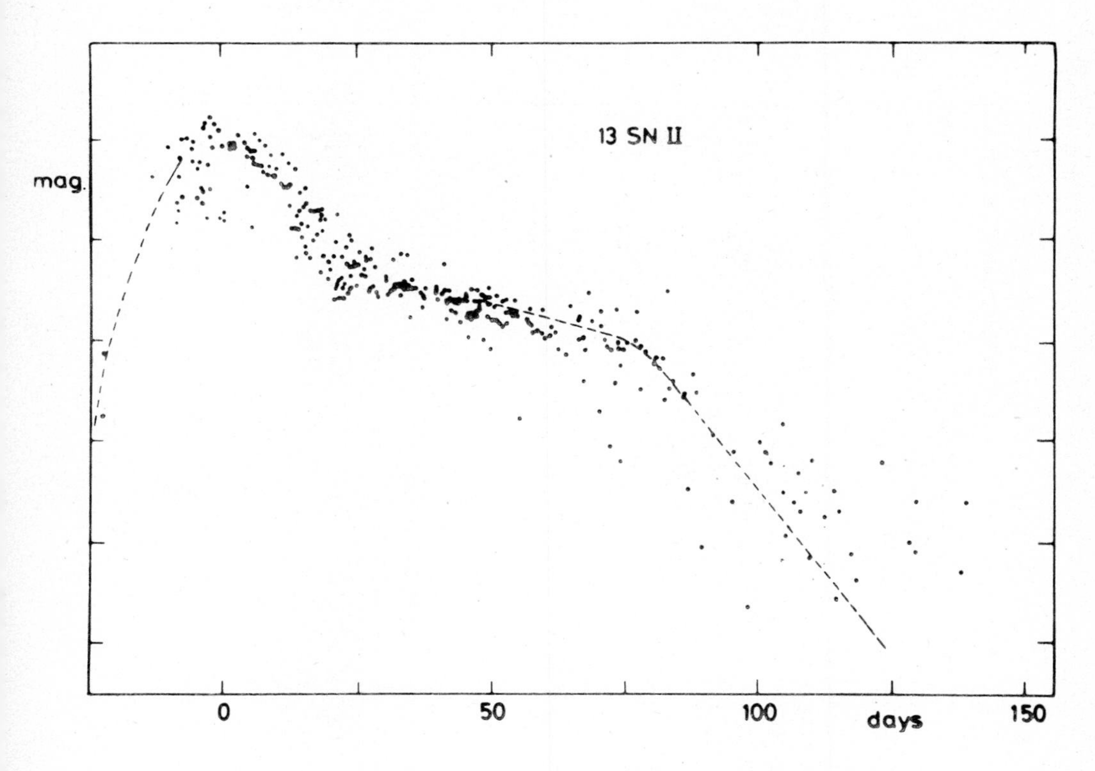

Fig. 1.7. Composite light curve obtained from 13 Type II supernovae. (From Barbon et al, 1974b).

X-ray wavelengths. However the radio remnants of supernovae can be detected on the remote side of the galaxy.

Over 120 extended galactic radio sources have been identified as SNRs. To date, no attempt to detect radio emission from extragalactic SNRs (other than 14 detected in the Magellanic Clouds) have produced conclusive results. We have therefore the anomaly of many extragalactic supernovae observed but no unambiguously identified extragalactic SNRs, and at the same time no galactic supernovae observed in modern times but many galactic SNRs.

Most of the catalogued galactic SNRs are probably so old that no records of the supernova explosion which created them can be expected to exist. For this reason, the few available historical observations of supernovae, although crude by present day standards, are very important. If a particular object can be identified with certainty as the remnant of a new star recorded in a particular year, then the time that the remnant has been developing is precisely known. Several such identifications

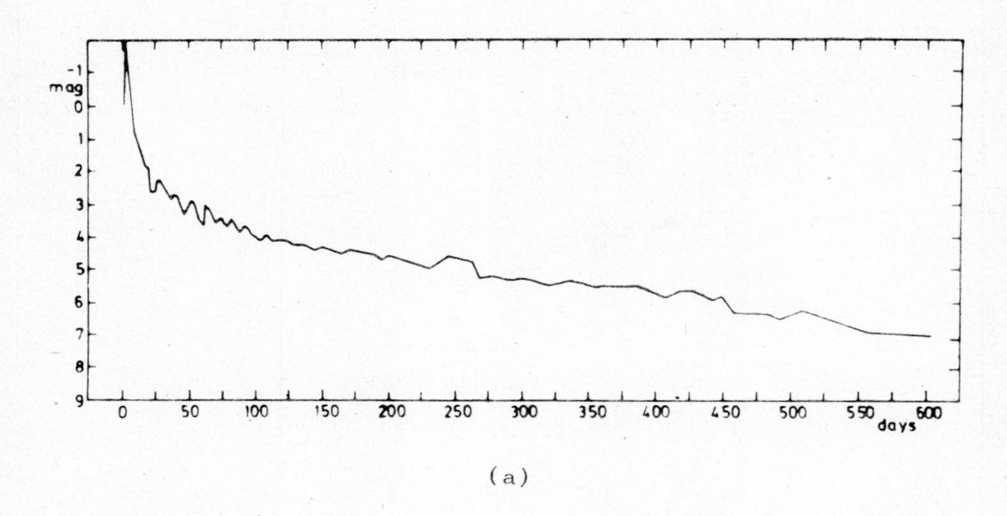

(a)

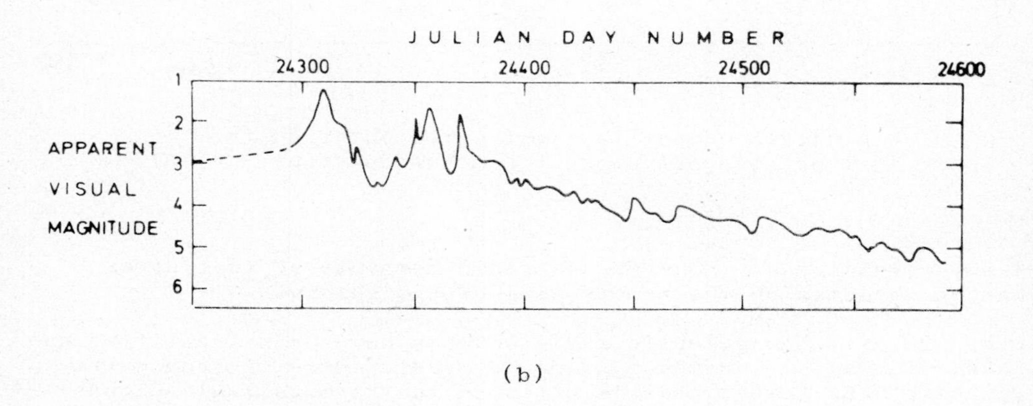

(b)

Fig. 1.8. (a) The light curve of a 'fast' nova, showing the typical rapid decrease in brightness after maximum - Nova Aquilae (1918), the brightest discovered since the invention of the telescope. (From Spencer Jones, 1961). (b) The light curve of a 'slow' nova, showing the typical irregular variation of brightness - Nova Pictoris (1925). (From Payne - Gaposchkin, 1957).

could then provide valuable observational evidence to test current theories on the evolution of supernova remnants, and on the nature of supernova explosions. Hence there can be little doubt that the historical astronomical records must be regarded as among the most valuable legacies which the ancient world has bequeathed to modern science.

Chapter 2

THE SEARCH FOR THE HISTORICAL RECORDS OF SUPERNOVAE

Usable historical records of new stars appear to be found in only four principal sources : Medieval European monastic chronicles; Arabic chronicles, astrological works etc.; post renaissance European scientific writings; and Far Eastern histories and diaries.

Possible historical sources of new star observations which might come readily to mind are the Greek and Roman Classics and the Late Babylonian astronomical texts. However, these turn out to be of little value. It is doubtful whether there are any useful references to new stars in the Classics. According to Pliny (Natural History, Book II) the appearance of a new star (*nova stella*) inspired Hipparchus to compile the first known star catalogue, giving co-ordinates of more than 1,000 stars. Pliny did not give a date for the appearance of the new star, but the date of Hipparchus' catalogue is known to be close to 130 BC (cf Dicks, 1960). As Pliny alludes to the motion of the star, there is a distinct likelihood that it was a comet. The report is of little interest to us here since no positional details are given, and in any case Pliny wrote two centuries after the event; he may merely be quoting hearsay. For a full discussion see Fotheringham (1919). Stothers (1976) has recently discussed possible references to new stars in Roman history. He is of the opinion that there are two allusions to a supernova in AD 185, but believes that evidence for other novae and supernovae is flimsy. We feel that even the references to the supernova of AD 185 are questionable, and in any case they give us no useful details.

The Late Babylonian astronomical texts, which cover the period from about 700 to 50 BC, contain large numbers of records of lunar and solar eclipses, occultations, etc. The principal collection of these tablets is in the British Museum. Plate 1 shows a small fragment which was originally part of an astronomical diary for the year 175 Seleucid (137 - 136 BC). On lines 13 to 15 can be found the most accurate description of a solar eclipse from ancient or medieval times (for a full translation see caption to Plate). We learn from Professor Abraham Sachs of Brown University, USA, who is a leading expert on the Late Babylonian texts, that there are a few observations of comets among the Babylonian inscriptions, but the texts which have been studied contain no references to novae or supernovae. This is possibly a result of the fragmentary condition of the tablets because bright novae, as well as supernovae, are rather rare events (see Chapter 3).

Plate 1. A Babylonian tablet referring to a solar eclipse in 136 BC. Lines 13 to 15 may be translated as follows (Sachs, personal communication) :
"Daytime of the 28th the north wind blew. Daytime of the 29th, 24 <u>uš</u> after sunrise, a solar eclipse beginning on the south west side ..... Venus, Mercury and the Normal Stars (i.e. the stars which were above the horizon) were visible; Jupiter and Mars, which were in their period of disappearance (i.e. between last and first visibility) were visible in that eclipse .... (the shadow) moved from south west to north east. (Time interval of) 35 <u>uš</u> for obscuration and clearing up (of the eclipse). In that eclipse north wind which ....."
The unit of time (<u>uš</u>) is the interval of time required for the heavens to turn through 1°.

Recently Brandt *et al.* (1976) have drawn attention to various cave paintings by American Indians which could possibly represent the supernova of AD 1054. Although this is an attractive hypothesis, the pictures tell us nothing about the star (see Chapter 8).

The first of the viable sources of historical records of new stars which we shall consider are the Medieval European monastic chronicles.

From about AD 1000 there was a flowering of literary activity in the numerous monasteries scattered throughout Europe; many monasteries began to keep chronicles of local events. These reports are often very detailed and make fascinating reading. Many of the earlier annals were published more than a century ago in the compilations *Monumenta Germaniae Historica* (Pertz, 1826→) and *Rerum Italicarum Scriptores* (Muratori, 1723→), both of which are well known to students of medieval European history. On the other hand, the numerous later chronicles commencing after about AD 1200 are relatively inaccessible.

The chronicles are in the main concerned with affairs of the monastery and local events. However, occasional reference is made to earthquakes and the more remarkable celestial phenomena, especially large solar eclipses and bright comets. Frequently an eclipse or comet is regarded as the major event of the year. (For translations of many of these records see Newton, 1972, and Muller and Stephenson, 1975). Occasionally annalists show a considerable interest in astronomical matters, recording aurorae, meteors and even occultations, as well as the more striking phenomena, but such curiosity is rare. As far as the appearance of a new star is concerned, scarcely less concern could have been shown. Either the chroniclers did not have adequate knowledge of the constellations to recognise a new star-like object, or they held such matters in little regard. To give an example of the former attitude, Simeon of Durham, a 12th century English chronicler, describes an occultation of a certain "bright star" by the totally eclipsed Moon. This occurred some four centuries before his time - the recorded date corresponds to AD 775 November 24 - but although the source of Simeon's account is unknown, the description is clearly that of an eyewitness (Arnold, 1885). Computation by the authors shows that the "bright star" was in fact the planet Jupiter, but it is evident that the observer was ignorant of its planetary nature. We might imagine that had a new star suddenly appeared in the sky, an observer of this calibre would not have recognised the fact, unless the star was extremely brilliant.

A particularly intriguing work is the continuation of the chronicle of Cosmas of Prague by a canon of the cathedral of Vysehrad in the city. This continuation, which covers the period AD 1126 to 1142, is published in volume 9 of *Monumenta Germaniae Historica*. Our author, who obviously had a keen interest in astronomy, and is writing as an eyewitness, records observations of solar and lunar eclipses, aurorae, meteors and parhelia (mock Suns). However, it is evident from his writings that he has no more than a hazy knowledge of the planetary motions, and cannot even identify the planet Venus with certainty. In the year AD 1131, the canon reports two strange stars, but gives only vague locations. He writes, "I do not believe that there are men who know which among them was called Lucifer" (i.e. Venus). In fact, it seems probable from the directions in which the "stars" appeared that each was Venus.

Before we criticise, we might well ask ourselves how many educated people today having no particular interest in astronomy can point out Jupiter or even Venus. Many will remember the close conjunction of these two planets which took place early in 1975. On the evening of February 17, the minimum separation was less than 0.2 deg. and the spectacle aroused considerable interest. However, it seems likely that even at the present time a solitary new star, unless it was unusually bright, would pass unnoticed to all but the persistent skywatcher.

There is a further, less obvious explanation why we find scarcely any references to new stars in the chronicles of medieval Europe. We have ample evidence, mainly from the Far East, that two bright supernovae were visible in the 11th century AD. The star of AD 1054 appeared in Taurus, and was thus well placed for northern observers. When it first appeared it would be prominent in the east before dawn. There are no known European records of it (see Chapter 8), possibly because it does not seem to have been much brighter than Venus. On the other hand, the star of AD 1006 was so brilliant that it could not have failed to attract widespread attention. It appeared in the southern constellation of Lupus (declination -37°), but despite its low altitude, it would be very prominent south of about latitude 48°N and even a casual observer could scarcely have failed to be impressed by it (see Chapter 7). Yet to our knowledge the only <u>certain</u> European records of the star are to be found in the chronicles of the monasteries of Benevento (41°N) and St. Gallen (47°N).

It seems that the Aristotelian concept of a perfect, changeless celestial vault was firmly rooted throughout Christendom until the Renaissance. Sarton, in a passage quoted by Needham (1959), the source of which is not specified, is particularly good on this. He writes, "The failure of medieval Europeans and Arabs to recognise such phenomena (i.e. new stars) was not due to any difficulty in seeing them, but to predjudice and spiritual inertia connected with the groundless belief in celestial perfection".

Whatever the explanation, the paucity of medieval European records of novae and supernovae is disappointing in the extreme, for the number of eclipse and cometary observations is enormous. Probably the best list of new stars in European chronicles is that of Newton (1972), which is compiled largely from <u>Monumenta Germaniae Historica</u>. However, the observations which he lists are of little scientific interest and reinforce what we have already said about the poor state of astronomical knowledge which prevailed in Europe at the time.

What seems at first sight a reasonably promising observation is recorded by the annalist Albertus in his chronicle of the monastery of Stade (Germany). This work is published in volume 16 of Monumenta Germaniae Historica. Albertus states that in the year AD 1245 (he is writing as a contemporary) a star appeared about May 4. He tells us that it was toward the south in Capricorn. It was large and bright, but red. It could not be Jupiter because Jupiter was then in Virgo. Many claimed that it was Mars because of its colour. After July 25 it was no longer bright and it continued to diminish day by day.

We have studied the observations with the aid of Tuckerman's (1964) planetary tables. It is evident that the "star" was, in fact, Mars for the planet was almost stationary in Capricorn from about May until September and was in opposition on July 21. Newton (1972) thought that the star was a nova and cited the gradual loss of brightness as evidence of this, although he emphasised the need for a check on the planetary positions before a definite judgement was reached. However, the fading was due to Mars receding from the Earth after opposition, although the indicated date seems a little early for noticeable loss of brightness.

To summarise, European records of possible new stars are of little value, and, with the exception of the observations made in AD 1006, are of historical interest only. Further research into the many unpublished chronicles seems likely to repay few dividends.

We now proceed to the Arabic writings.

The Arabs were certainly restricted in their outlook by the influence of Aristotle, but it was by no means a stranglehold, as in Europe. Medieval Arabian astronomy was far superior to its European counterpart. As the result of an extensive search through Arabic chronicles, treatises on astrology, etc., Goldstein (1965) was able to bring to light several detailed accounts of the new star of AD 1006 (see Chapter 7). We learn from Goldstein (personal communication) that there do not appear to be any extant Arabic records of the star of AD 1054. In the hope of making further progress, we have initiated a search of the literature of the period, but at the present time we can only express the hope that some useful results may emerge.

Before proceeding to the Far Eastern observations, which prove to be by far the most valuable, we shall briefly mention the Post Renaissance European observations.

As far as we are aware, no new star (i.e. nova or supernova) was recorded in Europe between AD 1006 and 1572. Two supernovae appeared shortly before Galileo first used a telescope in AD 1609 to study the heavens (AD 1572 and 1604), but no similar event has been observed in our own galaxy since then. It is with the advent of the telescope that our survey stops.

When the supernova of AD 1006 was seen, European astronomy was far behind that of the Orient. However, by the late 16th century the study of the stars had made great progress in Europe, whereas in the Far East it had remained virtually static. What can be described as truly scientific observations were made by Tycho Brahe and others in AD 1572, and these enable the position of the star to be fixed with some precision. Further, a remarkably accurate light curve can be drawn from the comparisons which Tycho made between the brightness of the supernova and that of other stars and planets (see Chapter 10). Kepler, Fabricius and others did much the same for the supernova of AD 1604 (see Chapter 11). As a result we know far more about these two stars than any which had previously appeared.

Both of these comparatively recent supernovae were observed in China and Korea (not, it seems, in Japan), but the reports from these countries are fairly unexceptional apart from the Korean observations of the AD 1604 event. However, whereas astronomy in Europe had just blossomed forth, the orientals had maintained a constant study of the heavens from earliest times. It is no coincidence that practically all of the new stars recorded in history were observed only in the Far East.

Because of the fundamental importance of the oriental records in the study of historical supernovae, it seems appropriate to give a brief outline of the history of the three countries China, Japan and Korea.

Oriental history began with China some time around 1500 BC. Prior to this time it is impossible to distinguish history from legend. The earliest historical dynasty is the Shang. According to orthodox Chinese history this is thought to have followed the Hsia Dynasty in 1766 BC, but although there are ample remains from the Shang, nothing has survived from the Hsia. Indeed, there is not a single reference to substantiate the existence of the Hsia Dynasty on the oracle bone fragments from the Shang.

Perhaps a satisfactory date for the beginning of the Shang Dynasty would be about 1500 BC. The people of Shang practised divination on a large scale using tortoise shells and the bones of various animals. After use, the bones were buried to prevent defilement. Towards the end of last century, vast quantities of these "dragon bones", as they were called, bearing a very primitive form of ideographic writing, were discovered near An-yang at the site of one of the main Shang capitals (Yin). As a result, we now have a very detailed knowledge of Shang divination, but still only a very incomplete picture of the history of the period. Because of their nature, these oracle texts contain very few astronomical records and these are mainly of eclipses (see Needham, 1959). Needham drew attention to two oracle bones which seemingly referred to the appearance of a new star near Antares and its subsequent fading (see Fig. 2.1). A translation of the illustrated text is given by Needham as follows :

Fig. 2.1. An oracle bone, seemingly referring to the appearance of a new star.

"On the 7th day of the month, a *chi-szŭ* day, a great new star appeared in company with Antares". Antares is described here as the "Fire (Star)", a name which is common in very much later texts. However, Jao Tsung-i (1955) suggests that this is not an astronomical observation, but a question to the ancestors about making offerings to the "Great Star" and to the "Fire (Star)".

The precise interpretation depends on how we read the character *hsin*, which could either mean "new" or "sacrifice". If the latter is correct, the "Great Star" could possibly be Venus or Sirius (we cannot be sure). Even if the former alternative is preferred, the uncertain duration makes the star scarcely worth considering as a supernova candidate (see Chapter 3). The first reliable observation of a new star was not made until more than a thousand years later.

The Shang Dynasty gave way to the Chou Dynasty about 1100 BC. This commenced almost immediately to establish a feudal system, which was the great strength of the dynasty in the first few centuries. However, after about 500 BC the individual feudal states became fully independent and the period known as that of the "Warring States" was ushered in. It was not until 221 BC that the empire was finally unified by Ch'in Shih-huang who became the first emperor of China. From this time onwards China has seen several partitions, but by and large the country has been a united empire under the rule of one man.

A list of the various Chinese dynasties is given in Table 2.1. In this table, the exact year given for the beginning of a dynasty may differ slightly from that given elsewhere. Much depends on the precise historical circumstances. Thus the Ch'in Dynasty was brought to an end in 206 BC by the adventurer Liu-pang, but, because of rivalry, it was not until 4 years later that he was proclaimed emperor (of the Han Dynasty). Again, the Yüan or Mongol Dynasty might be said to begin with the reign of Genghis Khan (AD 1206), but only in AD 1279 did the Mongols become absolute rulers of China when the Sung Dynasty was finally extinguished. In order to avoid somewhat disconcerting gaps in Table 2.1 we have normally taken the commencement of a dynasty as immediately following the previous one. The single exception concerns the Kin Dynasty. In AD 1141, the northern half of China was lost by the Sungs to the Kin Tartars. The two dynasties co-existed until AD 1234, when the Mongols overthrew the Kins.

Detailed history of China does not commence until the beginning of the (Former) Han Dynasty. The Chou Dynasty seems to have been extremely civilised, and presumably left numerous written records, but most of these were destroyed at the direction of Ch'in Shih-huang in 213 BC. On the advice of his prime minister, Li Szŭ, the emperor issued an order for the "Burning of the Books". This was a systematic seizure and burning of state records and any historical literature which did not concern the state of Ch'in. The object was to eradicate all memory of the former states which had vied with the emperor's own state of Ch'in for supremacy.

Ch'in Shih-huang was successful in destroying feudalism, but the people would not long tolerate such a totalitarian regime. Soon after his death in 209 BC, a series of rebellions culminated in the downfall of the Ch'in Dynasty and the establishing of the Han Dynasty, which apart from a short interregnum was to endure for more than four centuries. However, the holocaust which Ch'in Shih-huang instituted has left irreparable gaps in the history of ancient China. Presumably many further works perished in the flames when the Ch'in capital of Hsien-yang was destroyed in 206 BC. At any rate, little survives of Pre-Han history apart from the *Ch'un-ch'iu* (together with its enlargement, the *Tso-chuan*) and the *Shih-chi*. The *Ch'un-ch'iu* is the annals of the state of Lu, the birthplace of Confucius (hence its preservation), and covers the period 722 - 481 BC. The *Shih-chi* ("Historical Record") is a history of China from earliest mythical times down to 122 BC.

From the beginning of the Former Han Dynasty we possess remarkably detailed historical records. Much is due to Szŭ-ma Ch'ien (145 - 86 BC), who, in writing his monumental *Shih-chi*, set the pattern for all subsequent dynastic histories of China. It became standard practice for the official history of a dynasty to be compiled during the succeeding dynasty, usually by the Bureau of Historiography. Occasionally the compilation of a

Dynasty	Dates
Shang	c.1500 BC - c. 1100 BC
Chou	c.1100 BC - 481 BC
Chan-kuo (Warring States period)	481 BC - 221 BC
Ch'in	221 BC - 206 BC
Ch'ien-han (Former Han)	206 BC - AD 9
Hsin	AD 9 - 23
Hou-han (Later Han)	23 - 220
San-kuo (Three Kingdoms period)	220 - 265
Chin	265 - 420
Liu-sung	420 - 479
Nan-pei-ch'ao (Northern and Southern Dynasties)	479 - 581
Sui	581 - 618
T'ang	618 - 907
Wu-tai (Five Dynasties period)	907 - 960
Sung	960 - 1279
Chin (Tartar Dynasty)	1127 - 1234
Yüan (Mongol Dynasty)	1279 - 1368
Ming	1368 - 1644
Ch'ing (Manchu Dynasty)	1644 - 1911

Table 2.1. A schematic list of Chinese Dynasties

history might be undertaken centuries after the fall of the dynasty, but, generally speaking, the compilers had free access to all existing official records. The early invention of printing firstly using wooden blocks (possibly before AD 700) and later using movable type (from about AD 1050) has contributed enormously to the preservation of some 2,000 years of continuous history.

The 26 dynastic histories, together with their dates of writing (sometimes approximate) are given in Table 2.2. The source is Han Yu-shan (1955). A typical dynastic history contains four major divisions : Pên-chi ("Basic Annals") ; Piao ("Chronological Tables"), Chih ("Treatises"), and Lieh-chuan ("Biographies"). Of these, the Chih mainly concern us here since they are a fundamental source of astronomical records. The sources of later Chinese history are numerous, but in the search for astronomical records we can seldom do better than consult the dynastic histories. For this reason, in discussing Chinese historical sources we shall concentrate on the official histories, with a brief mention of the Shih-lu ("Veritable Records") which are available from some of the later dynasties.

Considering the four major divisions of a dynastic history, the Pên-chi are essentially annals of the emperors and are concerned with important affairs of state. Events are listed in chronological order, reign by reign. The term Piao, which may be translated literally as "to make manifest", is usually applied to the chronological tables. The name is taken from the object of these tables - to make manifest the essentials of the history.

The subjects of the Chih are diverse : rites, music, the calendar, astronomy, geography, law and punishments, the civil service - to name but a few. Frequently the Chih commences with the astronomical treatise, emphasising its fundamental importance, although this is by no means always the case.

By far the largest division of a history is the Lieh-chuan. The individual biographies are classified, e.g. scholars, officials, eminent women, artists, technicians. Such great emphasis on biography underlines the Chinese attitude to history. As Han Yu-shan (1955) puts it, "The worth of individuals and human achievements is considered crucial in the making of histories".

A word of caution : The dynastic histories present history in a unique and readily accessible form. However, the Chinese conception of history has its drawbacks. In reading a dynastic history one is always aware that what is written is what the official historian saw fit to preserve. In particular, it is clear that many of the astronomical records are mere summaries of the original observations with considerable loss of detail. However, we must not be over critical. The present authors have found the Chinese dynastic histories a seemingly inexhaustible goldmine of astronomical observations, without equal anywhere in the world.

	History	Period	Written	
1	Shih-chi	(Down to 122 BC)	104 - 87 BC	+
2	Han-shu	Former Han Dynasty	AD 58 - 76	+
3	Hou-han-shu	Later Han Dynasty	398 - 445	
4	San-kuo-chih	Three Kingdoms period	285 - 297	
5	Chin-shu	Chin Dynasty	644	*
6	Sung-shu	Liu-sung Dynasty	492 - 493	+
7	Nan-ch'i-shu		489 - 537	
8	Liang-shu	Northern	628 - 635	+
9	Ch'en-shu		622 - 629	+
10	Wei-shu	and	551 - 554	+
11	Pei-ch'i-shu	Southern	627 - 636	+
12	Chou-shu		c. 630	+
13	Nan-shih	Dynasties	630 - 650	
14	Pei-shih		630 - 650	
15	Sui-shu	Sui Dynasty	629 - 636	*
16	Chiu-t'ang-shu	T'ang Dynasty	940 - 945	*
17	Hsin-t'ang-shu	"	1043 - 1060	
18	Chiu-wu-tai-shih	Five Dynasties period	973 - 974	*
19	Hsin-wu-tai-shih	"	1044 - 1060	
20	Sung-shih	Sung Dynasty	1343 - 1345	*
21	Liao-shih	Liao Dynasty	1343 - 1344	*
22	Kin-shih	Kin Dynasty	1343 - 1344	*
23	Yüan-shih	Yuan Dynasty	c. 1370	*
24	Hsin-yüan-shih	"	1890 - 1920	
25	Ming-shih	Ming Dynasty	1672 - 1755	*
26	Ch'ing-shih-kao	Ch'ing Dynasty	1914 - 1927	*

Table 2.2. The twenty-six Chinese Dynastic Histories

* Compiled by the Bureau

+ Compiled by Official Decree

We have mentioned in passing the Shih-lu or "Veritable Records". These were written up at the end of each reign and were based on the Ch'i-chü-chu ("Diaries of Activity and Repose" of the emperor). The Ming-shih-lu and the Ch'ing-shih-lu still exist in their entirety, but these contain very little of astronomical interest. In this respect they are in marked contrast to their Korean equivalent, the Yijo Sillok (see below).

It is necessary to explain why astronomical observation was practised on such a large scale. From earliest times a "Grand Historian" (Ta-shih) was appointed. According to Han Yu-shan (1955), the title Ta-shih is to be found among the Shang oracle bone inscriptions. During the Chou Dynasty, and possibly during the Shang, the Grand Historian had a double duty to perform. Since natural phenomena were regarded as closely related to human affairs, the historian recorded both natural happenings and human events. In particular, he performed the duties of astronomer and astrologer. During the Former Han Dynasty, by which time affairs of state had become more complex, the concurrent duties of the historian were delegated and observation of natural phenomena was assigned to a separate office, the T'ai-ch'ang ("Office of Sacrificial Worship"). From this time onwards, the "Astronomical Bureau", as it is usually termed, continued in operation throughout the rise and fall of dynasties until the establishment of the modern republic (AD 1912). The officials, in general, were hereditary, a factor which no doubt made the system resistant to change.

One of the functions of the Astronomical Bureau was the maintenance of the calendar. No doubt this was the main reason for making the numerous eclipse observations which are recorded in the various astronomical treatises. However, the astrological interpretations which accompany many of the eclipse records, and the vast number of observations of no calendar importance (e.g. occultations, planetary conjunctions, sunspots, comets, novae, aurorae) for which astrological prognostications are also supplied, make it clear that divination was the principal object of the Bureau. Ho Peng Yoke (1969) has made a detailed study of the Astronomical Bureau in Ming China. Commenting on an account in the Ming-shih, he remarks, "With the help of the Bureau Staff they (i.e. the Director and Deputy Director of the Bureau) observed or took measurements on the Sun, the Moon, the stars and asterisms, winds and clouds, and the colour of vapours, and submitted confidential reports to the Emperors whenever there were abnormalities". Needham (1959) quotes a fascinating account of the workings of the Ch'ing astronomical bureau as the Jesuit Lecomte found it in AD 1696:
"They still continue their observations. Five Mathematicians spend every night on the Tower in watching what passes overhead; one is gazing towards the Zenith, another to the East, a third to the West, the fourth turns his eyes Southwards, and a fifth Northwards, that nothing of what happens in the four Corners of the World may scape their diligent Observation. They take notice of the Winds, the Rain, the Air, of unusual Phenomenas, such as are Eclipses, the Conjunction or Opposition of Planets, Fires, Meteors, and all that may be useful. This they keep a strict

Accompt of, which they bring in every Morning to the Surveyor of the Mathematicks, to be registered in his Office".

In essence, this state of affairs had remained virtually unchanged since at least Han times. Because of its importance, the Imperial Observatory was normally built adjoining the Palace (see Fig. 2.2). As the capital was changed, (usually at the beginning of a new dynasty), a new observatory would be built. There is good reason to believe that practically all of the astronomical observations recorded in the various astronomical treatises were made at the capital. The workings of the Astronomical Bureau seem to have been a closely guarded secret (cf. Needham, Wang-ling and Price, 1960). Because of the pre-eminent position of the Bureau, falsification of astronomical observations for astrological purposes was possible, and indeed probable. How else can we explain the following list of "observations" recorded in a Korean work, the *T'aejŏng Sillok* ("Chronicle of the Reign of King T'aejŏng "). The year corresponds to AD 1406. (We have converted all dates to the Julian Calendar.)

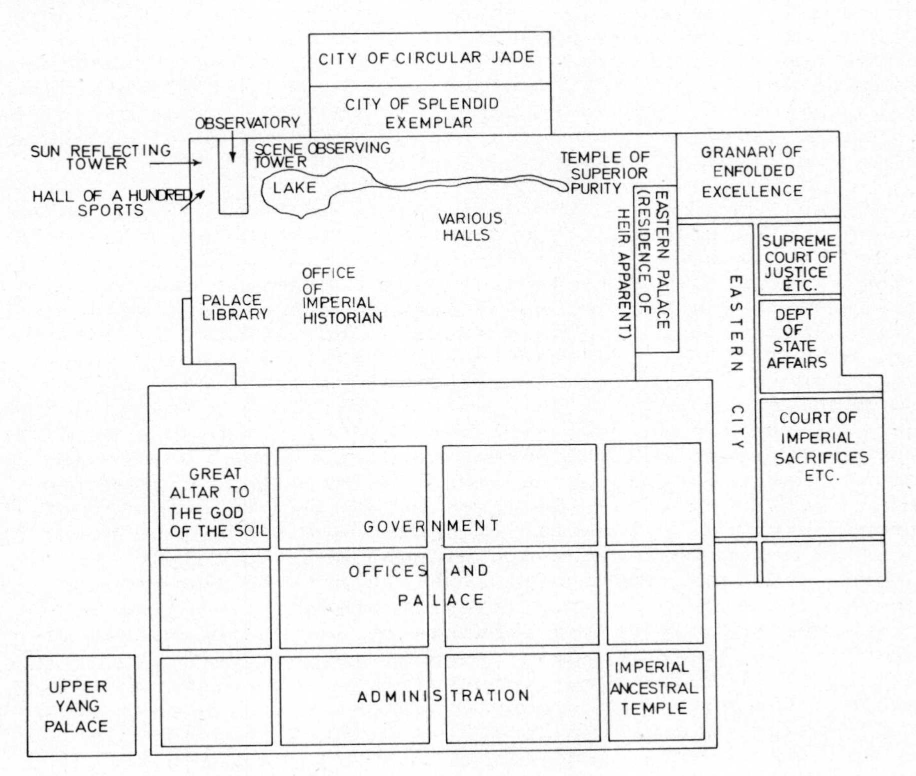

Fig. 2.2. Plan of the T'ang capital of Lo-yang.

August 10. "The Moon invaded Mars".
August 14. "The Moon invaded Mars".
August 17. "The Moon invaded Mars; Jupiter and Mars invaded one another".
August 21. "Mars invaded Jupiter".
August 25. "The Moon invaded Jupiter".

Computation shows that only the last two dates are correct. It seems that the King and court officials had virtually no knowledge of astronomy. This is by far the most glaring example of fabrication that we are aware of, but others have been noted (cf. Bielenstein, 1950).

The astronomical treatises in the dynastic histories show that almost every conceivable kind of observation that could be made with the unaided eye was reported by the Bureau. Most treatises adopt a detailed classification of observations. Thus the Chin-shu treatise contains 9 sub-sections : "thunder-like noises"; eclipses of the Sun; solar haloes and sunspots; lunar changes (eclipses, haloes, etc.); planetary conjunctions; daylight appearances and conjunctions of the Moon and five planets with asterisms; "Ominous stars" and "guest stars" (comets and novae); meteors; and "extraordinary clouds and vapours" (usually aurorae).

There are frequent references to the more spectacular astronomical phenomena (especially eclipses) in the Pên-chi, or Imperial Annals, but these are couched in layman's terms and are practically devoid of descriptive details. Occasionally biographies (e.g. of astronomers) mention astronomical events, but this is rare. The astronomical treatises prove to be our most profitable source.

The first astronomical records from Japan date from the 7th century AD. At this time there was no permanent capital, but on the death of each emperor, the capital was transferred to a new site. The various capitals were all in the Nara Plain, near Kyōto. By AD 710, the first regular capital was established at Nara. This was modelled on the residence of the T'ang emperor at Ch'ang-an. In the preceding centuries Chinese and Korean cultural influences had been profound. Particularly important, from about AD 400 onwards the Chinese system of writing had been introduced by way of Korea. Practically all of Japanese astronomical records which we have studied are written in Classical Chinese.

By the end of the 8th century (AD 794), the capital was moved to Heian (later known as Kyōto). This was again modelled on the Chinese capital of Ch'ang-an, and remained the residence of the Japanese emperors until AD 1869. Towards the end of the 12th century, military governors took control of the country. While the emperor continued to live at Kyōto, the seat of the Shogun, or "generalissimo" was Kamakura, near the site of present-day Tōkyō.

There is no history of Japan comparable with the dynastic histories of China. The astronomical records are in general scattered in a variety of works, such as diaries of courtiers and privately written histories. A valuable source is the Dainihonshi ("History of Great Japan"), compiled in AD 1715. This contains a section devoted to astronomical records. However, the study of Japanese astronomical observations owes much to the patient research of Kanda Shigeru, who published an exhaustive compilation in 1935. Kanda made a thorough search through histories, diaries, temple records, etc., and it would be very difficult to improve on his work. In the present book we have used Kanda as an index to all our sources of Japanese records, although in almost every case we have consulted the original works which Kanda cites.

The Chinese system of naming the constellations (see below) and Chinese astronomical terms were in use in Japan certainly by the 8th century AD. Generally speaking, Japanese records are similar in style to those of China. Once more, the motive was mainly astrological, but prognostications referred to Japan rather than the country of origin of the system. Particularly after about AD 800, astronomical records are quite frequent, but at no time do we find the degree of regularity or observational accuracy achieved in China.

Korean history (as distinct from legend) may be conveniently subdivided into four distinct periods: the epoch of the Three Kingdoms; the Silla ascendancy; the Wang Dynasty; and the Yi Dynasty. Around 50 BC, largely under Chinese influence, the various tribes of the peninsula amalgamated into three independent kingdoms - Silla in the south-east, Paekche in the south-west, and Koguryŏ in the north, bordering with China. It was probably about this time that the Chinese script was introduced, but it was not until the 4th century AD that this became widely disseminated. The three states were almost continually at war with one another and with China and Japan in the early centuries.

Around AD 700, in alliance with T'ang China, Silla conquered Paekche and Koguryŏ. However, some surviving Koguryŏ retainers established a new dynasty called the Parhae near the Chinese border. Thus the Korean peninsula and Southern Manchuria were divided between the two powers - Silla in the south and Parhae in the north. The gradual decline of Silla in the 9th century led to the establishment of a new kingdom named Koryŏ in AD 918. Wang Kŏn, its first king (who later took the title of T'aejo), successfully unified the whole country for the first time in AD 936. His capital was Sŏngdo (Kaesŏng). The Wang Dynasty was to endure for more than three centuries- until AD 1392 when General Yi Sŏng-gye overthrew the regime and established the Yi Dynasty with its capital at Han-yang (Sŏul). The country then took the modern name of Chosŏn. The Yi Dynasty lasted until the Japanese annexation in AD 1910.

The Samguk Sagi ("History of the Three Kingdoms"), which covers the period until the commencement of the Wang Dynasty was compiled in AD 1145 by order of King Injŏng of Koryŏ. It contains a few astronomical records but these are of doubtful reliability. One solitary astronomical structure exists from the Three Kingdoms period. This is in Kyŏngju, the former capital of Silla, and was built in AD 647 during the reign of Queen Sŏngdok of Silla. The observatory, illustrated in Plate 2, is called Ch'ŏmsŏng-dae, literally "Tower for Star Observation". Its purpose has recently been questioned by Kim Yong-woon (1974), but the interpretation of the structure as an observatory still seems most likely. If so, it is the oldest existing observatory in the Orient.

Coming to the history of Koryŏ, the Koryŏ-sa ("History of the Kingdom of Koryŏ") is an authoritative work patterned on a Chinese dynastic history. It was compiled in the early decades of the Yi Dynasty by official decree and was finally completed in AD 1451 after many revisions. This work, which is by far the most valuable history of the period, contains an astronomical treatise of three chapters. From this it is clear that after about AD 1,000 a high level of attainment was reached. As in Japan, Chinese astronomy was adopted, and indeed the astronomical treatise of the Koryŏ-sa is indistinguishable from a similar treatise in the Chinese dynastic history.

The most valuable source of astronomical records during the Yi dynasty is the Yijo Sillok. Sillok is the equivalent of the Shih-lu in China, i.e. "Veritable Records" written up at the end of each reign (see above). However, the Korean work is very much more detailed than its Chinese counterparts. In the Sillok, astronomical records are not contained in a separate section, but are included in chronological order along with affairs of court and state. Fortunately, summaries of the various observations are classified in the Chŭngbo Munhŏn Pigo, a late Korean historical summary, which acts as a useful index to the records in the Sillok.

It might be pointed out that all four Korean works mentioned are written in Classical Chinese. The modern Korean alphabet, known as Han-gŭl, and consisting of 11 vowels and 28 consonants was invented by King Sejŏng in the 15th century AD. This achieved considerable popularity, but met with opposition from scholars, hence its neglect in literary works.

We shall conclude this chapter with discussions on the Chinese calendar and the Chinese conception of the constellations.

The Chinese calendar, which was adopted from an early period in both Korea and Japan, is fundamentally luni-solar. However, an independent system of counting days makes reduction to the Western calendar a relatively simple operation. In China, years were originally numbered from the beginning of the reign of a king (before the Unification in 221 BC) or emperor (after the Unification). Emperor Wên, of the Former Han Dynasty (179 - 156 BC) was the first to establish reign periods, i.e. subdivisions of a reign. There were usually two or three such periods in a reign, although the exact number could vary from

Plate 2. The "Tower for Star Observations" at Kyŏngju. Built in AD 647, it is the oldest existing observatory in the Orient.

only one to more than ten. From the 8th century AD onwards, Japanese emperors adopted a similar system, but it was never introduced in Korea. It might be mentioned here that the ruler of Korea was always known as a king, never an emperor. Tables of reigns and reign periods are readily available, e.g. Moule and Yetts (1957) or Tsui Chi (1947) for China and Tchang (1905) for Japan and Korea.

Probably from as early as the Shang Dynasty in China, alternate months of 29 and 30 days, with occasional adjustments, enabled the calendar to keep pace with the Moon (the mean length of the synodic month - from one new Moon to the next - is 29.5306 days). Certainly by the Chou Dynasty (if not before), an astronomical new Moon, rather than an observed crescent, began the month, and as a result all but a few solar eclipses are recorded on the first day of the lunar month. Most years contained 12 months. The problem of the incompatibility of the synodic month and the tropical year (365.2422 days) was surmounted by the insertion of an intercalary month every $2\frac{1}{2}$ years or so. From Han times, intercalation was very regular the object being to make the new year (beginning of the first month) fall on the second new Moon after the winter solstice.

Side by side with the luni-solar calendar, an independent system of day numbers is in operation. This 60 - day cycle, similar in many respects to our week, was in use during Shang times. The cycle was formed by the combination of a series of 10 symbols - the "celestial stems" (*t'ien-kan*) - with another series of 12 symbols - the "terrestrial branches" (*ti-chih*). By taking the symbols in pairs (one from each series) in cyclical order, 60 possible permutations are obtained.

The 10 *t'ien-kan* are as follows : *chia*, *i*, *ping*, *ting*, *wu*, *chi*, *kêng*, *hsin*, *jen*, *kuei*. The 12 *ti-chih* are : *tzŭ*, *ch'ou*, *yin*, *mao*, *ch'en*, *szŭ*, *wu*, *wei*, *shen*, *yu* *hsü*, *hai*. We have shown the full sexagesimal cycle in Table 2.3. To give an example of the use of this table, the day *kêng-ch'en* is the 17th day of the cycle. It should be noted that many combinations cannot occur, e.g. there is no day *jen-wei*.

Analysis of the eclipse records in the *Ch'un-ch'iu* or "Spring and Autumn Annals" (722 BC - 480 BC) demonstrates that the cycle has been unbroken since this period (see Muller and Stephenson, 1975). There is really no reason for suspecting that any interruption occurred during the Shang or early Chou dynasties, so that it seems probable that the cycle has had a continuous existence for more than three millennia. So important has the sexagesimal cycle proved that in oriental records the day of the month, unless it happens to be the first or last day is seldom given (except in Japan), whereas the cyclical day is almost always specified. The same system was in use for numbering years, but it never achieved the popularity of the day-numbering system. It is interesting to note that the 12 *ti-chih* are used for naming the double hours. The first double hour (*tzŭ*)is the period between 11 p.m. and 1.a.m. local time, and so on throughout the day. Thus *wu* corresponds to

	tzu	ch'ou	yin	mao	ch'en	szu	wu	wei	shen	yu	hsu	hai
chia	1		51		41		31		21		11	
i		2		52		42		32		22		12
ping	13		3		53		43		33		23	
ting		14		4		54		44		34		24
wu	25		15		5		55		45		35	
chi		26		16		6		56		46		36
keng	37		27		17		7		57		47	
hsin		38		28		18		8		58		48
jen	49		39		29		19		9		59	
kuei		50		40		30		20		10		60

Table 2.3. Days of the Sexagesimal Cycle

9 a.m. to 11 a.m.

Tables produced by Hsüeh Chung-san and Ou-yang I (1956) enable the rapid conversion of Chinese dates to the Western calendar. (Julian or Gregorian) for any date between AD 1 and 2000. These tables are equally useful for converting Japanese or Korean dates when used in conjunction with a list of reigns or reign periods. Numerous examples of dates expressed in the oriental style are given in succeeding chapters.

Some time before the Han Dynasty (just when is impossible to say owing to the paucity of extant records), an amazingly complex astronomical / astrological system evolved in China. The sky was subdivided into several hundred small asterisms, each containing an average of 5 or 6 stars. Around AD 275 the Astronomer Royal estimated from early star maps then in existence that there were 283 star groups and 1464 stars (Ho Peng Yoke, 1966). Later estimates vary considerably, but these figures are fairly representative.

The names of the various asterisms span the whole spectrum of life in China, e.g. Ti-wang ("Emperor"), T'ai-tzŭ ("Crown Prince"), Shang-shu ("Secretaries"), Huan-chê ("Court Eunuchs"), T'ien-chuang ("Celestial Bed"), Ch'uan-shê ("Guest Houses"), Nei-ch'u ("Inner Kitchen"), Sha-hsing ("Execution Star"), T'ien-chiang ("Celestial River") to name but a few, chosen more or less at random. A complete list is given by Ho Peng Yoke (1966).

Celestial events, such as a planet, comet or new star entering a particular asterism, or the occurrence of an eclipse, were regarded as precursors of terrestrial events involving the corresponding personage, object, state, etc. The astronomical treatises of the earlier dynastic histories (particularly those of the Han and Chin) contain detailed prognostications for almost every observation. We have selected the following examples from the Hou-han-shu and the Chin-shu, referring to the solar eclipse of AD 65 and the new star of AD 369 respectively :
(AD 65). "8th year (of the Yung-p'ing reign period), 10th month, (day) wu-wu, the last day of the month. There was an eclipse of the Sun and it was total. It was 11 degrees in (Nan-) tou. (Nan-) tou represents the state of Wu. Kuang-ling, as far as the constellation are concerned, belongs to Wu. Two years later Ching, King of Kuang-ling, was accused of plotting rebellion and committed suicide".
(AD 369). "4th year of the T'ai-ho reign period of Hai-hsi. 2nd month. A guest star was seen at the western wall of Tzŭ-wei. When we come to the 7th month it finally disappeared. The interpretation when a guest star guards Tzŭ-wei is assassination of the Emperor by his subjects. In the 6th month Huan-wei dethroned the Emperor who became the "Duke of Hui-hsi" .

It can be seen from these two examples that the delay between the occurrence of an omen and its acknowledged fulfilment was usually comfortably long. Two years was fairly typical, evidently to ensure the maximum chance of success.

According to Ho Peng Yoke (1966), the belief in astral influence on state events germinated in the *Ch'un-ch'iu* period (722 - 480 BC), and became firmly implanted through the teaching of Liu-hsiang (77 to 6 BC), a well known astronomer and mathematician. In any event, it was to endure for more than two millennia, and established itself in both Korea and Japan. Certain emperors were opposed to astrology and divination, e.g. Wu-ti, who in AD 268 passed an edict to ban the study of these subjects, and Yang-ti, who about AD 610 ordered all books dealing with these subjects to be burnt. However, and fortunately for modern science, these enlightened rulers were in the minority. Ironically enough, the astronomical treatises in the *Sung-shu* and *Chin-shu* regard an observation of Mars in AD 287 as an omen of Wu-ti's death; evidently his influence was short-lived!

The prognostications accompanying the observations seem trivial now, but it is a direct result of the importance attached to astrology that we possess such a vast number of observations of all kinds, and in particular such an impressive list of new star sightings. Further, since the substance of a particular prognostication depended very much on which asterism the new star appeared in, Far Eastern records are usually careful to give a fairly accurate location. Normally only the asterism is named, which might mean that the star appeared anywhere within an area of say 100 square degrees of sky, but occasional positions were measured to the nearest degree by the Korean astronomers (see Chapter 11). Needham (1959) believes that from the Han onwards measurements of position were made in degrees, but the historiographers simplified the information they received in writing the official histories. This may well be the case.

Oriental positions are often expressed in terms of the 28 *hsiu* or "lunar mansions". Each lunar mansion covers a range of right ascensions, obviously averaging about 13 deg. in extent, but the individual extensions are very variable. The *hsiu* are named after asterisms near the celestial equator, and the boundary of each *hsiu* is fixed by the location of a certain "determinant star" of the asterism. The lunar mansions are in order (with the determinant stars in parentheses) :

1. *Chüeh* (α Vir.)
2. *K'ang* (κ Vir.)
3. *Ti* (α^2 Lib.)
4. *Fang* (π Sco.)
5. *Hsin* (σ Sco.)
6. *Wei* (μ^1 Sco.)
7. *Chi* (γ Sgr.)
8. *Nan-tou* (φ Sgr.)
9. *Niu* (β Cap.)
10. *Hsü-nü* (ε Aqr.)
11. *Hsü* (β Aqr.)
12. *Wei* (α Aqr.)
13. *Ying-shih* (α Peg.)
14. *Tung-pi* (γ Peg.)
15. *K'uei* (η And.)
16. *Lou* (β Ari.)
17. *Wei* (41 Ari.)
18. *Mao* (η Tau.)
19. *Pi* (ε Tau.)
20. *Tsui-hsi* (λ Ori.)
21. *Shen* (ζ Ori.)
22. *Tung-ching* (μ Gem.)
23. *Yü-kuei* (θ Cnc.)
24. *Liu* (δ Hya.)
25. *Ch'i-hsing* (α Hya.)
26. *Chang* (ε^1 Hya.)
27. *I* (α Crt.)
28. *Chen* (γ Crv.)

The asterisms listed above lie roughly on a great circle. At the present time, the declinations vary between about -30 and +30 deg., but around 2500 BC, the circle encompassing the lunar mansions corresponded reasonably closely to the celestial equator. There have been attempts to use this method in tracing the origin of the lunar mansions, but in the absence of confirmatory historical evidence, we can no more than hypothesise (for a discussion see Needham, 1959).

Lunar mansions are frequently used to specify the positions of new stars. Fortunately, in such instances oriental records are quite clear to distinguish whether the actual asterism is referred to or the range of right ascensions bearing its name. When the name of the asterism alone is used, the star group itself is intended. However, when the character hsiu (i.e. "lunar mansion") is appended, or a star is described as so many degrees (tu) within the asterism, the right ascension is implied.

An excellent representation of the oriental asterisms is shown in Plates 3 and 4. These maps were drawn by Jesuit astronomers in China in AD 1747 (the charts were originally published by Rigge, 1915). The two hemispheres are shown separately in polar projection. On the maps, the radial lines represent the boundaries of the 28 lunar mansions. The inner circle on the Northern Hemisphere chart depicts the boundary of the region of perpetual visibility from about 40° N (the latitude of Peking). The corresponding circle on the Southern Hemisphere chart is the edge of the south polar region, invisible from North China. Evidently the asterisms within this latter region were drawn from Western observations made in the Southern Hemisphere. It will be noticed that no attempt is made to distinguish between stars of different magnitude. This feature is characteristic of practically all oriental star maps.

In our discussion of the new star observations we have made use of a variety of oriental star maps in addition to the above. A list of these is as follows :

1. An early Jesuit map (date around AD 1600) preserved in the Consiglio Nazionale delle Ricerche at Bologna, Italy. Our attention was first drawn to this by Professor G. Palumbo. The chart is similar in form to the later Jesuit map discussed above.
2. A series of maps in the Ku-chin-t'u-shu-ch'i-cheng, a Chinese encyclopedia dating from about AD 1725, but containing pre-Jesuit star maps from more than a century earlier. These maps are crudely drawn, but they are particularly valuable since they distinguish three different types of star "magnitude" - an almost unique feature.
3. A similar series of late pre-Jesuit charts in another encyclopedia, the San Tsai Tu Hwei. Here no attempt is made to represent star brightness.
4. A Korean planisphere of AD 1395. This is an exact copy of a very early Chinese star map. The original stone engraving was presented to the King of Koguryŏ by the Emperor of China at an early period. This stone was lost in a river when Koguryŏ fell to Silla in AD 672. However, a new engraving based on a preserved rubbing of the original, was made in AD 1395. This, and

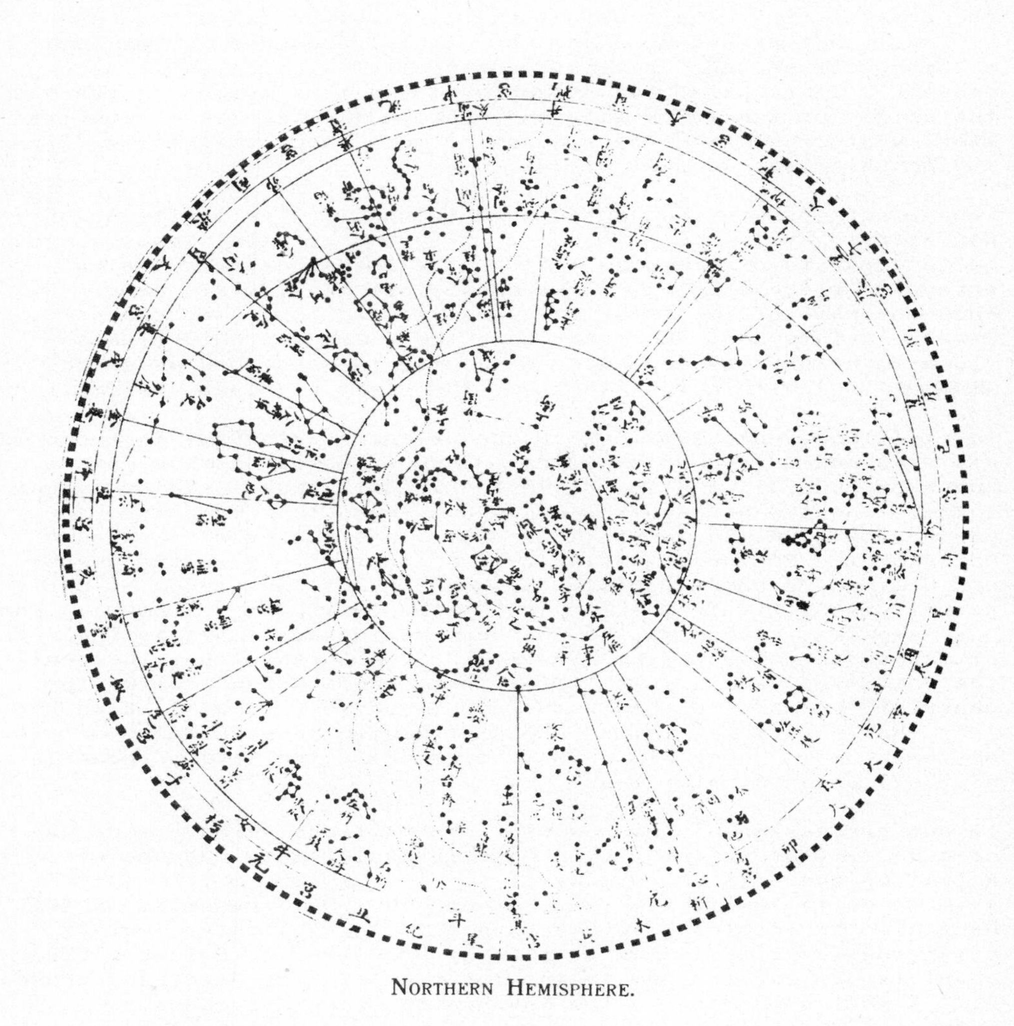

Plate 3. A Chinese star chart, drawn by Jesuits in Peking in the 18th century, shows asterisms in the northern celestial hemisphere.

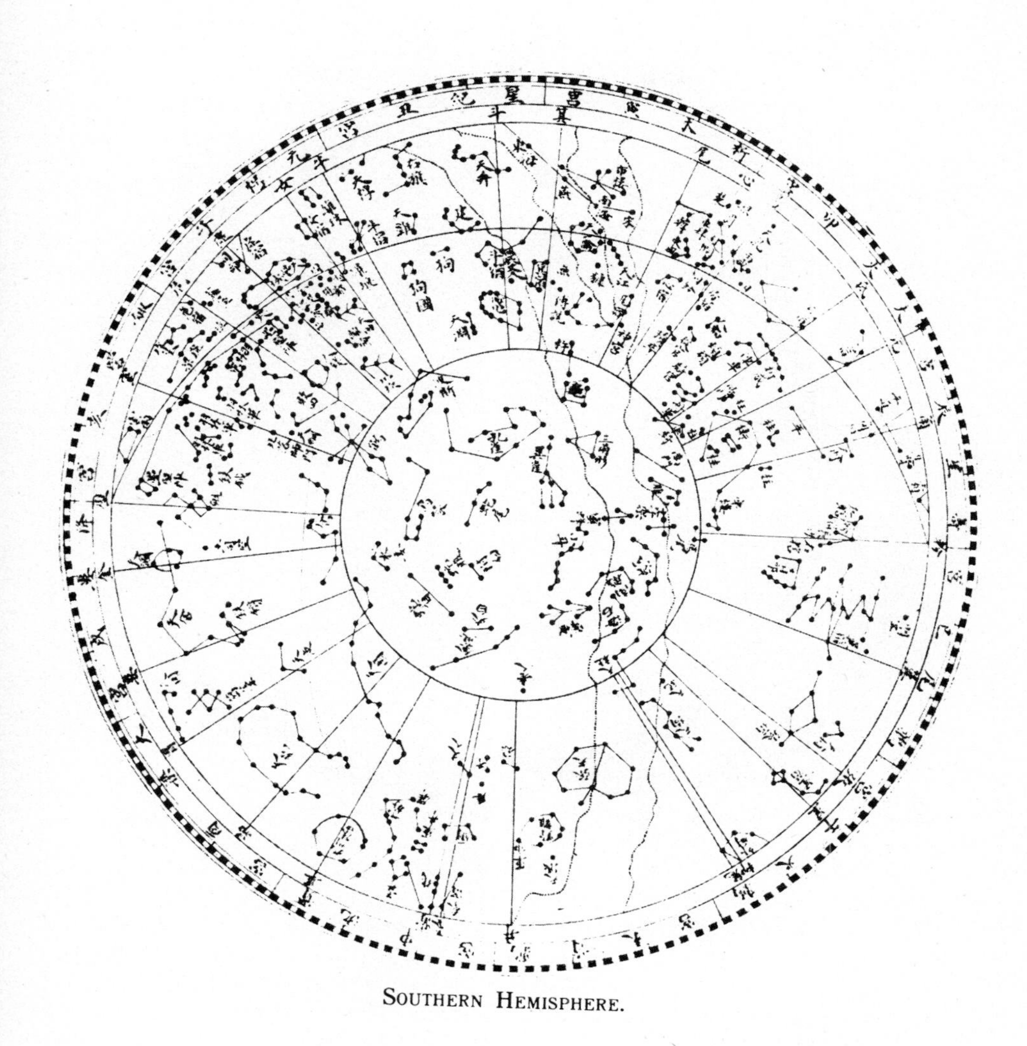

SOUTHERN HEMISPHERE.

Plate 4. A companion star chart to Plate 3, showing the southern celestial hemisphere.

Plate 5. Bronze planisphere found some time last century in a Japanese junk. This shows the asterisms visible from latitude 35° N. The planisphere, which is about 35 cm. (14 in) in diameter, is an almost exact copy of a Korean star map engraved on stone in AD 1395 and still surviving. This was itself copied from a very early Chinese map. The position of the equinox corresponds to about 100 BC, so that the original map is of very ancient origin. The circles represent the celestial equator, ecliptic and 'Purple Palace' (circle of perpetual visibility). Other features are the boundaries of the 28 'lunar mansions', represented by radial lines, and the Milky Way, shown as a curving irregular band.

a later copy (AD 1668) still survive in a Sŏul museum. Rufus (1913), who made a detailed study of the planisphere, estimated the epoch when the chart was first prepared as close to 100 BC from the position of the equinox and celestial pole. The Royal Scottish Museum, Edinburgh, contains a bronze planisphere which was found in the 1870's in a Japanese junk which was wrecked on an island off the coast of Japan. Rufus and Chao (1944) drew attention to the fact that this planisphere is either a careful copy of the Korean map or shares a common origin with it. The Japanese planisphere is illustrated in Plate 5. Idealization of certain star groups is evident, but the original Chinese astrography was apparently still a valuable aid to navigation some two millennia later.

5. The Soochow chart prepared in AD 1193 and engraved on stone in AD 1247. This map is rather idealized and several of the asterisms appear to be displaced to allow space for their names. An excellent rubbing is illustrated by Rufus and T'ien Hsing-chih (1945).

6. A fairly accurate series of maps contained in the *Hsin-i-hsiang-fa-yao*, a work by Su Sung and dating from AD 1092. These are probably the earliest printed star maps (cf Needham, 1959).

7. The crude star maps which form part of a vast collection of manuscripts which Sir Aurel Stein discovered last century in Tunhuang (Kansu). The date of these maps is AD 940 (Needham, 1959).

8. *Hsing-ching* ("Star Manual"). This work, which is contained in the Taoist Canon, is probably a compilation of Pre-Han and Han catalogues with later additions up to about AD 600.

Having considered the sources of the historical records, we are now ready to proceed to a discussion of the new star records themselves.

Chapter 3

A CLASSIFICATION OF THE FAR-EASTERN NEW STARS

Basically, three kinds of new star were recognised in the Far East. These are *k'o-hsing* ("guest stars" or "visiting stars"), *po-hsing* ("rayed stars" or "bushy stars") and *hui-hsing* ("broom stars" or "sweeping stars"). These will be considered in turn.

K'o-hsing (which will be subsequently abbreviated to *k'o*) seems to have been the general term to describe a new star-like object. The well known new stars of AD 1006, 1054, 1572 and 1604 were identified in this way and we might thus expect *k'o* to be synonomous with novae and supernovae. On the other hand, there are frequent references to moving *k'o* throughout oriental history (more than 20 are catalogued by Ho Peng Yoke, 1962), so that usage of the term must be treated with caution. The nucleus of a comet resembles a star, so that if no tail is evident confusion seems possible.

Po-hsing (*po*) is the standard term to describe an apparently tailless comet. In the astronomical treatise of the *Chin-shu* (the official history of the Chin Dynasty) we find the following:

"By definition a comet pointing towards one particular direction is a Hui Comet and one that sends out its rays evenly in all directions is a Po comet".

Ho Peng Yoke (1966) attempted to explain a *po* as a tailed comet seen almost head-on, i.e. in opposition to the Sun, but this can be the explanation for only a small proportion of such stars. The records of *po* indicate that they could appear at any elongation from the Sun, and furthermore were frequently visible for several months. In the catalogue of Ho Peng Yoke (1962) the motions of some 10 *po* and *k'o* for three months or more are described. Reading the various descriptions of these objects, it is difficult to avoid the conclusion that some comets can remain visible for an extended period without developing an appreciable tail. We occasionally find accounts of *po* which seem to imply that the star remained stationary for a number of days - i.e. there is no reference to motion. The possibility thus exists that some of these stars were novae rather than comets. However, it should be remembered that most oriental records which are available to us are mere summaries of the original observations, possibly with considerable loss of detail. As with *k'o* some care must be exercised in classifying *po* in modern terms.

Hui-hsing (*hui*) is the general term for a tailed comet. In the same section of the *Chin-shu* as the previous definition we find the following description of a *hui*:

"Its body is a sort of star, while the tail resembles a broom. Small comets measure several inches in length, but the larger ones extend across the entire heavens".

Since the use of the term hui implies that the star had a definite tail, the probability of a nova or supernova being described in this way is obviously very low. Many records of hui give an estimate of the length of the tail, frequently several tens of degrees, and a detailed account of the motion is often given. On the other hand, often a brief mention of the sighting of a hui, without any reference to motion, is reported. Here again we may be dealing with a brief summary of the original observation.

A particularly remarkable feature of oriental observations of comets is the extent to which two independent reports of the same star can vary. To give a random example, an account of the motion of Halley's Comet in the Sung-shih (the official history of the Sung Dynasty in China) is extremely detailed, whereas only a single sighting is recording in the Koryŏ-sa (the official history of the Wang Dynasty in Korea). Numerous similar examples could be quoted. (The year of the above sighting is AD 1066.)

It has been pointed out by Ho Peng Yoke (1962) that the supernova of AD 1572 was called a hui in the Ming-shih-kao, the draft version of the official history of the Ming Dynasty. However, this error was corrected in the history itself (the Ming-shih), where the star is described as a k'o.

As an alternative to hui-hsing, the term ch'ang-hsing ("long star") is not infrequently used. Once again this denotes a tailed comet, and we occasionally read of ch'ang-hsing with tails stretching across the sky (for instance, in 135 BC and AD 128). As for hui, we feel that, except when there seems to be very good reason to the contrary, all ch'ang-hsing should be regarded as comets.

Finally, a few miscellaneous "stars", usually nameless, are from time to time reported in Far Eastern historical sources. In such instances it seems best to treat each individual case on its own merits.

The first catalogue of temporary stars recorded in history, compiled for its own sake, was probably that of Fujiwara Sadaie, a Japanese poet-courtier. Following the appearance of a k'o in AD 1230 (this was in reality a comet - see below), Fujiwara made a search of Japanese historical works for references to new stars. In his diary, the Meigetsuki ("Diary of the Full Moon"), he gave details for eight such stars ranging in date from AD 642 to 1181. All are to be found in the entry for AD 1230 December 13. This diary is an important source of Japanese records of novae and supernovae.

In China at about the same time, Ma Tuan-lin in his encyclopedia, the Wên-hsien-t'ung-kao, included a list of temporary stars in his section on Chinese astronomical observations. However, this list, although very useful to modern astronomers, did not have a special motive. In the 18th and 19th centuries a number of lists of historical new stars were compiled - Pingré (1783), Biot (1843), Humboldt (1851) and Williams (1871) - however, these are of little value at the present day. Perhaps the first person to realise the significance of temporary stars other than comets was Lundmark (1921). He produced a catalogue of 60 suspected novae and supernovae based on earlier secondary sources, in which he deduced approximate co-ordinates and plotted the galactic distribution of the stars. Unfortunately, Lundmark gives us very little information on his criteria for classifying stars as novae rather than comets. His catalogue has been superseded by modern ones which are based on direct consultation of original sources.

Kanda Shigeru (1935) made an extensive compilation of Japanese astronomical observations recorded in a wide variety of sources. This covers the period up to AD 1600, and includes a list of new stars (both comets and novae). Kanda's work forms a valuable source of references.

The first modern catalogue of possible novae and supernovae is due to Hsi Tsê-tsung (1955). This contains ninety entries up to AD 1690, mainly obtained from Chinese and Japanese historical materials. Complete accounts and references are given for each star, together with estimated co-ordinates, and the galactic distribution is shown diagrammatically. Hsi Tsê-tsung criticised several of Lundmark's selections, but his own catalogue is far from reliable. _Po_ and _hui_ are included along with _k'o_, almost at random. Again, certain stars are suggested as supernovae without foundation, while pairs of stars are linked as possible recurrent novae even though the individual positions were only approximately recorded.

The Smithsonian translation of Hsi Tsê-tsung's catalogue is excellent. However, when using this work it should be pointed out that _po-hsing_ is translated as "sparkling star".

More recently, in collaboration with Po Shu-jen, Hsi Tsê-tsung (1965) produced an improved version of his previous catalogue. Korean sources and a Vietnamese history were also consulted, and again a total of ninety stars up to AD 1690 were listed. However, this work suffers from much the same defects as its predecessor, and like it should be used with caution. It might be remarked that the abridged translation by Yang (1966) is far superior to the, at times, shockingly bad NASA translation.

The brief catalogue of Chu Sun-il (1968) contains summaries of 28 Korean observations of temporary stars up to AD 1665, together with estimated positions. However, Chu does not distinguish between the various types of star which he lists and it is thus impossible to assess the reliability of any particular sighting without consulting the original record (or a translation published elsewhere).

Of the fundamental catalogues, we are left with the two compilations of Ho Peng Yoke (1962, 1970). Both works are entirely based on Far Eastern sources, the original giving Chinese, Japanese and Korean observations up to AD 1600 and the supplement Chinese observations from AD 1368 to 1911. The great value of Ho Peng Yoke's work lies mainly in the impartial nature of his compilation. He set out to compile a list of comets _and_ novae, and it is left to the reader to decide for himself the nature of any particular object. This is backed up by an exhaustive search of the literature and excellent translations. Further, the oriental star maps which he provides in his original catalogue, although not always reliable (Stephenson, 1971), enable an estimate of the uncertainty in the position of any particular star to be made.

Recently, Pskovskii (1972) has produced a secondary catalogue of possible novae and supernovae based on the various modern fundamental catalogues discussed above. Justifiably, Pskovskii criticises the somewhat naive approach of some of these works, but his own methods of selection are open to criticism. For instance, he includes without justification many _po_ in his catalogues which were rejected by Hsi Tsê-tsung and Po Shu-jen (1965). In particular he states that "no comet has ever been observed with the unaided eye for more than four months", implying that all stars of longer duration were novae or supernovae. In fact, Ho Peng Yoke (1970) gives full details of the motions of several comets which were observed in the Far East for longer than four months, up to a maximum of six and a half months.

The catalogue of pre-telescopic novae and supernovae to be presented here makes no claim to originality in so far as the material used is concerned. It is based on a collation of the fundamental catalogues discussed above. However, in view of the authors' considerable experience in handling oriental astronomical records, it is hoped that the present work is as objective as possible. It should be remarked that we have independently consulted almost all of the original records of the most important observations.

The only new stars sighted in the West which are included in the present catalogue are those of AD 1006, 1572, and 1604. An Arabic report of the new star of AD 1006, falsely dated as AD 827 by Humboldt (1851) and others, was correctly dated by Goldstein (1965); see Chapter 7. Excluded from the catalogue are the new stars which are supposed to have appeared in AD 945 and 1246. Our only authority for these is the 16th century astronomer Cyprianus Leovitius. Leovitius did not name his sources, and the original records have never been uncovered. Humboldt drew attention to a new star which a certain Cuspinianus was supposed to have observed in AD 393. According to Humboldt, this appeared close to α Aquilae and rivalled Venus. However, Lynn (1884) pointed out that Cuspinianus was a 16th century AD writer, and he traced the original report to the 5th century _Chronicon_ of Marcellinus. This states, "A star rising from the north at cockcrow, and burning like Venus rather than shining, appeared, and on the 23rd day it disappeared." The approximate date is the beginning of September in AD 389. Only a very rough location is given, and Lynn emphasized that "no star near the constellation of Aquila could have risen at cockcrowing" around this time of year. He in fact concluded that the star was identical with a comet described by the contemporary writer Philostorgius, who gives an account of its motion. So much for the reliability of secondary sources!

Coming now to the Far East, the bulk of the new stars which we have catalogued were originally classified as *k'o*. These include stars of both long and short duration, as well as those of indefinite duration for which there is only a mention of the first sighting. We have listed all *k'o* for which a celestial position is given and for which there *is* no definite mention of a tail or motion. It is unfortunate that many reports of *k'o* give only the general direction (usually one of the four cardinal directions) in which the star appeared and this, of course, is useless. Deciding whether or not a tail is referred to is not always straightforward. Frequently a record may merely state that a star "measured x *ch'ih*". Actually, *ch'ih* ("feet") is a linear unit, but Kiang (1972) and Stephenson (1971) have independently shown that, roughly at least, *ch'ih* and *tu* (degrees of 365.25 to a circle) were synonomous. To the average eye, a bright star appears to have "rays" of an appreciable extent due to spherical aberration, astigmatism, etc. In order to reduce the number of spurious objects, we have systematically omitted all stars which were said to measure more than 1 *ch'ih* (roughly twice the Moon's apparent diameter), unless an unusually long duration, with no apparent motion, is reported.

We have excluded all isolated sightings of *po*. Use of this term implies that the star possessed a definite form, and when no duration is recorded the probability of it being a nova seems low. It is likely that a number of *k'o* of indefinite duration were comets, but here at least the "correct" term for a nova is applied. Where a *po* is recorded with a *definite* duration, but without any hint of motion or tail, there is a reasonable possibility that the star was misidentified, and accordingly we have included such examples in our catalogue.

With two exceptions (*hui* in 5 BC and AD 247) all *hui* and *ch'ang-hsing* are excluded from the catalogue. These two stars remained visible for many days and there is no hint of any motion. Further investigation thus seems desirable. All other such stars for which no motion or tail is mentioned were visible for about 25 days or less (usually much less) and accordingly seem scarcely worth considering as possible novae or supernovae.

It should be pointed out that two well known *k'o* are omitted from our catalogue. These are the supposed supernovae of AD 902 (Pskovskii, 1963) and AD 1230 (Shklovsky, 1968). The former was considered in some detail by Stephenson (1975). The Chinese record of the star (in the *Hsin-t'ang-shu*, the "New Book of the

T'ang Dynasty") distinctly states that it changed position - "During the first month of the second year of the T'ien-fu reign period, (11 February to 12 March AD 902) a guest star like a peach was at *Tzŭ-wei* beneath (i.e. to the north of) *Hua-kai*. It gradually moved and reached *Yű-nu*. On the day *ting-mao* (2 March) a meteor (*liu-hsing*) left *Wên-chang* and reached the guest star; the guest star did not move. On the day *chi-szŭ* (4 March) the guest star was at *K'ang* and guarded it. In the following year (*ming-nien*) it still had not faded away".

The whole record is most difficult to interpret. We are told quite specifically that the star moved, so that a tail-less comet seems to be the obvious interpretation. Further, all the details relate to the first month and only at the end do we find that it was still visible the following year. This is much too long a duration for a comet. We feel that we must read *jih* ("day") for *nien* ("year"), although the characters are very different, so that the last sentence commences, "On the following day". All the events then occurred in the same month and the latter part of the entry could be merely interpreted as a commentary to the effect that although the meteor seemed to come in contact with the guest star it did not disturb it in any way. On the other hand we may be dealing with two independent guest stars, but it is impossible to be definite. Regretfully we must abandon the AD 902 star as of uncertain nature.

Ho Peng Yoke (1962) showed quite definitely that the star of AD 1230, which was said to be visible for 4 months (in the *Sung-shih*, the official history of the Sung Dynasty) was a comet, by quoting a parallel record in the *Kin-shih* (the official history of the Kin Tartar Dynasty) which gives a detailed description of the motion of the star. It should be noted that both histories describe the star as a *hui*, the standard term for a comet. Only in Japan was it called a *k'o*. This example, in which a comet which was visible for several months, and yet could be recorded in one history without any mention of motion, suggests that other similar instances might occur.

Although the AD 902 and AD 1230 new stars are excluded from our catalogue, all other stars which have at one-time been seriously proposed as supernova candidates are included.

Coming now to the few references to other types of star, we have included in our catalogue all stars not in the above categories for which there is no certain reference to motion or a tail, regardless of duration. Here again, it is highly likely that some comets are included, but this is unavoidable with such sketchy material.

Generally speaking, it is obvious that the reliability of the observations included in the catalogue varies enormously. In an attempt to allow for this we have devised a somewhat arbitrary classification scheme. This will be discussed below.

Table 3.1. Catalogue of pretelescopic galactic novae and supernovae.

REF. NO.	DATE	PLACE	TYPE	DURATION	CLASS	RA (1950.0)	DEC (1950.0)	l	b
01	BC 532 Spring	C	'star'	-	5	20^h 50^m	-10°	40°	-30°
02	204 Aug./Sep.	C	po	10 days	5	14 20	+20	20	+70
03	134 Jun./Jul.	C	k'o	-	4	16 00	-25	350	+20
04	77 Oct./Nov.	C	k'o	-	4	11 10	+75	130	+40
05	76 May /Jun.	C	chu	-	5	1 40	+25	135	-35
06	48 May	C	k'o	-	4	18 40	+25	10	-10
07	47 Jun./Jul.	C	k'o	-	4	4 00	+65	140	+10
08	5 Mar./Apr.	C	hui	70 + days	2	20 20	-15	30	-25
09	AD 61 Sep. 27	C	k'o	70 days	2	14 10	+35	60	+70
10	64 May 3	C	k'o	75 days	2	12 20	- 5	290	+55
11	70 Dec./Jan.	C	k'o	48 days	1	9 40	+25	215	+45
12	85 Jun. 1	K	k'o	-	5	-	+65	-	-
13	101 Dec. 30	C	k'o	-	4	9 40	+25	215	+45
14	107 Sep. 13	C	k'o	-	4	6 30	+10	200	0
15	125 Dec./Jan.	C	k'o	-	4	17 10	+10	30	+25
16	126 Mar. 23	C	k'o	-	5	12 00	+10	270	+70
17	185 Dec. 7	C	k'o	20 months	1	14 20	-60	315	0
18	222 Nov. 4	C	k'o	-	4	12 30	0	290	+60
19	247 Jan. 16	C	hui	156 days	2	12 30	-20	295	+40
20	290 Apr./May	C	k'o	-	4	-	+65	-	-

(1)	(2)	(3)	(4)	(5)	(6)	(7)	(8)	(9)	(10)
21	304 Jun./Jul.	C	k'o	-	4	4 20	+15	180	-25
22	329 Aug./Sep.	C	po	23 days	5	12 30	+55	130	+65
23	369 Mar./Apr.	C	k'o	5 months	1	-	+65	-	-
24	386 Apr./May	C	k'o	3 months	1	18 30	-25	10	-10
25	393 Feb./Mar.	C	k'o	8 months	1	17 10	-40	345	0
26	396 Jul./Aug.	C	'star'	50 + days	2	4 00	+20	175	-25
27	402 Nov./Dec.	C	k'o	2 months	2	11 10	+10	240	+60
28	421 Jan./Feb.	C	k'o	-	4	11 30	-15	275	+45
29	437 Jan. 26	C	'star'	-	5	6 40	+20	195	+ 5
30	483 Nov./Dec.	C	k'o	-	5	5 30	0	205	-15
31	537 Jan./Feb.	C	k'o	-	4	-	+65	-	-
32	541 Feb./Mar.	C	k'o	-	4	-	+65	-	-
33	561 Sep. 26	C	k'o	-	4	11 30	-15	275	+45
34	641 Aug. 6	C	po	25 days	5	12 20	+20	265	+80
35	684 Dec./Jan	J	po	2 weeks	5	3 40	+25	165	-25
36	722 Aug. 19	J	k'o	5 days	3	1 00	+60	125	0
37	829 Nov.	C	k'o	-	4	7 50	+15	205	+20
38	837 Apr. 29	C	k'o	22 days	3	7 00	+10	205	+ 5
39	837 May 3	C	k'o	75 days	1	12 10	+ 5	280	+65
40	837 Jun. 26	C	k'o	-	5	18 00	-25	5	0

(1)	(2)	(3)	(4)	(5)	(6)	(7)	(8)	(9)	(10)
41	877 Feb. 11	J	k'o	-	4	23 50	+20	105	-40
42	891 May 12	J	k'o	-	4	16 50	-20	0	+15
43	900 Feb./Mar.	C	k'o	-	5	17 00	+10	30	+30
44	911 May/ Jun.	C	k'o	-	4	17 10	+15	35	+30
45	1006 Apr. 3	A,C,E,J	k'o	2 + years	1	15 10	-40	330	+15
46	1011 Feb. 8	C	k'o	-	4	19 20	-30	10	-20
47	1035 Jan. 15	C	'star'	-	5	1 20	+5	140	-55
48	1054 Jul. 4	C,J	k'o	22 months	1	5 40	+20	190	- 5
49	1065 Sep. 11	C	k'o	-	4	9 20	-25	255	+20
50	1069 Jul. 12	C	k'o	11 days	3	18 10	-35	0	-10
51	1070 Dec. 25	C	k'o	-	4	2 40	+ 5	165	-50
52	1073 Oct. 9	K	k'o	-	4	0 10	+ 5	105	-55
53	1074 Aug. 19	K	k'o	-	4	0 10	+ 5	105	-55
54	1138 Jun./Jul.	C	k'o	-	4	1 50	+20	140	-40
55	1139 Mar. 23	C	k'o	-	4	14 10	-10	335	+50
56	1163 Aug. 10	K	k'o	-	4	17 30	-20	5	+ 5
57	1175 Aug. 10	C	po	5 days	5	15 40	+50	80	+50
58	1181 Aug. 6	C,J	k'o	185 days	1	1 30	+65	130	0
59	1203 Jul. 28	C	k'o	9 days	3	17 10	-40	345	0
60	1224 Jul. 11	C	k'o	-	4	17 10	-40	345	0

(1)	(2)	(3)	(4)	(5)	(6)	(7)	(8)	(9)	(10)
61	1240 Aug. 17	C	k'o	-	4	17 10	-40	345	0
62	1356 May 3	K	k'o	-	4	5 50	+30	180	0
63	1388 Mar. 29	C	'star'	-	5	0 10	+20	110	-40
64	1399 Jan. 5	K	k'o	-	4	18 50	-20	15	-10
65	1404 Nov. 14	C	'star'	-	5	19 50	+30	65	0
66	1430 Sep. 9	C	'star'	26 days	4	7 30	+ 5	215	+10
67	1431 Jan. 4	C	'star'	15 days	4	4 50	-10	210	-30
68	1437 Mar. 11	K	k'o	14 days	3	16 50	-40	345	0
69	1460 Feb./Mar.	V	'star'	-	5	11 30	-15	275	+45
70	1572 Nov. 8	C,E,K	k'o	16 months	1	0 20	+65	120	0
71	1584 Jul. 11	C	'star'	-	5	16 00	-25	350	+20
72	1592 Nov. 28	K	k'o	15 months	1	1 20	-10	150	-70
73	1592 Nov. 30	K	k'o	4 months	1	0 50	+60	125	0
74	1592 Dec. 4	K	k'o	3 months	1	0 00	+60	115	0
75	1604 Oct. 8	C,E,K	k'o	12 months	1	17 30	-20	5	+ 5

Our revised catalogue of pre-telescopic galactic novae and supernovae which forms Table 3.1 contains information for seventy-five candidates observed from earliest times until the beginning of the telescopic era (AD 1609). The columns give in order:

1. A reference number.
2. The Julian or Gregorian calendar date (the Gregorian calendar is used from AD 1582 September 15, the date of its inception). Frequently only the month is given.
3. An initial letter denoting the place of observation (Arab Lands, China, Europe, Japan, Korea or Vietnam).
4. The type of star (k'o, po, hui, etc).
5. The duration of visibility, where known.
6. A classification number from 1 to 5, depending upon the reliability.

7 & 8. The approximate R.A. and dec. for epoch 1950.0, estimated from the oriental star maps of Ho Peng Yoke (1962, 1966).

9 & 10. The approximate galactic co-ordinates (l and b).

We have not considered it worthwhile to include the names of the asterisms in or near which each star appeared. This information is of minor interest to the astrophysicist, whereas the sinologist would require further details.

Referring to Table 3.1, it is curious that no durations between about twenty-five and fifty days are reported. This provides a means of selecting the most promising supernova candidates. As shown by Clark and Stephenson (1976), any star of duration less than about forty days can be virtually ruled out as a supernova. This is based partly on the evidence that there are only two possible references to the long period variable star Mira Ceti in Far Eastern history (AD 1070 and 1592) before its discovery in Europe by David Fabricius in AD 1596. This star has a period of roughly eleven months. At maximum it reaches a magnitude of between +2 and +4 and at minimum descends to about +9. It is thus only visible to the unaided eye for about half of the time and yet at a bright maximum it is the brightest star in its neighbourhood. At a declination of -3° it is very well placed for general observation. Taking conservative estimates of magnitude +3 as the minimum brightness for detection and +5.5 as the lower limit of unaided eye visibility, judging from the Type I and Type II light curves of Barbon et al (1974a,b) a supernova would take at least forty days, and possibly much longer, to decline over this range of brightness. There is the possibility that a supernova might appear shortly before conjunction with the Sun and thus be visible for only a few days. However in Table 3.1 only a single k'o of specified but short duration appeared near the sun. This occurred in AD 1069, and as discussed in Chapter 4 there is no known supernova remnant in its vicinity.

From the above argument, we should expect to find any supernovae among the longer duration stars (more than about fifty days), while stars of short duration (less than about twenty-five days) are not worth seriously considering as supernovae. This, of course, is not to imply that all stars of longer duration are supernovae; we might expect to find both slow novae and comets among this material. However, there can be little doubt that the short duration objects are a mixture of novae and comets only.

Referring to the classification in column 5 of Table 3.1, stars in classes 1 and 2 are all objects which were seen for more than about fifty days. Class 1 objects are all k'o for which there is no hint of any tail or motion in the record. Stars in class 2 are either downgraded k'o (because of possible reference to a tail or motion) or stars of other type. Basically class 3 objects are k'o of short but definite duration (up to a maximum of about twenty-five days). Class 4 stars are k'o of indefinite duration or unspecified stars of known duration. Finally, objects in class 5 are unspecified stars of indefinite duration or po of known duration. An object is downgraded by 1 class if there is any suggesting of a tail or motion.

It seems likely that most of the stars in class 3 were novae. However, the proportion of comets in classes 4 and 5 may be very significant.

Fig. 3.1 shows the positions of all of the stars in the catalogue in terms of galactic co-ordinates (l and b). The blank zone in the lower right of the figure represents the area currently not visible from 35°N latitude (the approximate location of the various Chinese, Japanese and Korean capitals). Stars of medium to long duration (more than about fifty days) are represented by shaded circles, others by open circles.

Outside of the blank area, the distribution in both l and b is fairly isotropic. This, of course, is what we would expect for comets. From the statistics of Galactic novae collected by Payne-Gaposchkin (1957), these stars in general show a marked concentration towards the galactic equator. The distribution of novae in galactic longitude is also very anisotropic, with a pronounced peak in the direction of the galactic centre ($l = 0^{\circ}$). However, it is rather problematical to assess the expected distribution of novae bright enough to attract the attention of the unaided eye. An indirect method is once more to make use of the almost complete lack of pre-1596 sightings of Mira Ceti. The maximum brightness of this star is about magnitude +2. A star brighter than +1 could scarcely be missed under normal circumstances. It would therefore appear that to stand a reasonable chance of being discovered by the early Far Eastern astronomers a maximum brightness of somewhere between these two values would be required. It seems appropriate to take the mean, i.e. +1.5.

Assuming an average magnitude of -7.5 for galactic novae (Payne-Gaposchkin, 1957) and allowing for interstellar absorption of approximately 1 mag. per kpc (a smaller value would be appropriate for stars well off the plane) gives an average distance for novae bright enough to be noticed of about 500 pc. At such

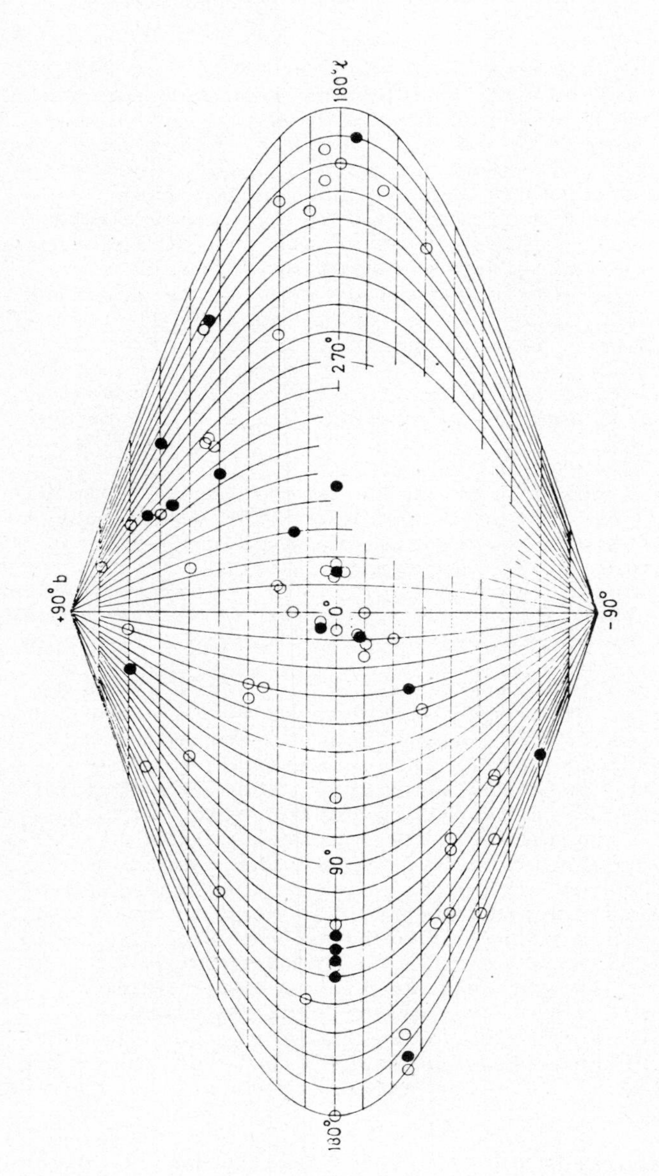

Fig. 3.1. Galactic distribution of pretelescopic novae and supernovae.

a small mean distance we would expect the distribution of historical novae in galactic longitude to be almost uniform.

Again from Payne-Gaposchkin (1957), the average distance of novae from the galactic plane in the vicinity of the Sun is about 275pc. A fairly isotropic distribution in galactic latitude is thus also expected for historically recorded novae.

It is obviously not possible to give a useful estimate of the proportion of novae in the catalogue. All that can be done is to select the most probable supernovae and novae and leave the nature of the remaining objects in doubt. Because of their vast distance from us (the nearest historically observed supernova, which occurred in AD 1006, lies at a distance of 1 kpc), supernovae tend to lie close to the galactic equator. From the investigation by Clark and Stephenson (1976), the mean value of b for young (less than 2,000 years old) nearby (within about 10 kpc of the Sun) SNRs is only $2^{\circ}.3$. Independent evidence for supernovae belonging to the plane of a parent galaxy has been produced by Vettolani and Zamorani (1976) from an interpretation of the distribution of 132 supernovae detected in spiral and irregular galaxies. We thus expect to find galactic supernovae only among the new stars of long duration <u>and</u> low galactic latitude, (see also Chapter 4).

Of the temporary stars of short duration, only two are sure novae. The records of both the stars of AD 837 (A) and 1437 imply that the objects remained fixed for several days. The latter appeared close to the galactic equator, but the duration of fourteen days is much too short to make it worth considering as a possible supernova, particularly as the star was more than 90 degrees from the Sun and would thus be visible in a dark sky for several hours each night. Presumably it was a fast nova. The star of AD 1431 is a possible nova of the DQ Her. type (with a deep minimum) but the record is very brief.

Attention has sometimes been drawn to the possibility of two seemingly independent new stars representing separate outbursts of the same (recurrent) nova. In view of the poor positioning of such stars in general and, with a few notable exceptions, the marginal evidence that any particular star was a nova, no suggested recurrent novae among the pretelescopic objects can be taken seriously.

The twenty stars of duration in excess of about fifty days are summarized in Table 3.2, and their galactic distribution is shown in Fig. 3.2. The columns of Table 3.2 given in order are:

1. The reference number in Table 3.1
2. The year of occurrence.
3. The duration of visibility.
4. The galactic latitude.

Of the new stars in high galactic latitude ($|b| > 25^{\circ}$), the five objects of 5 BC, AD 61, 64, 247, and 402 may well have been

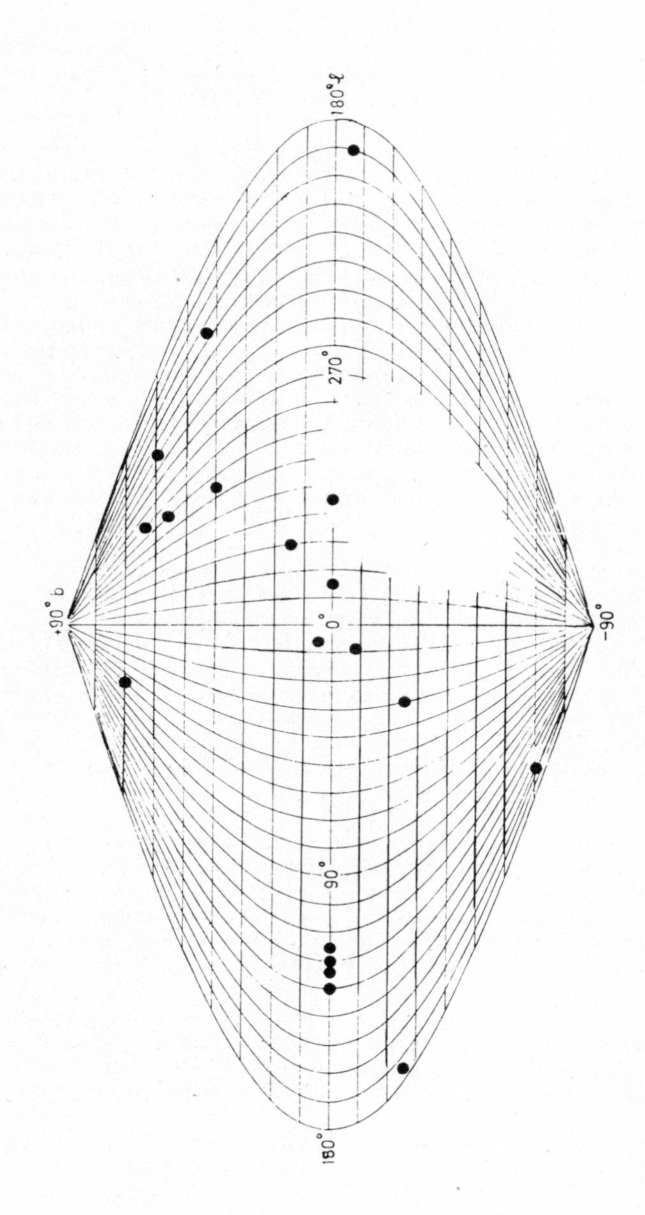

Fig. 3.2. Galactic distribution of pretelescopic new stars of long duration.

REF. NO.	YEAR	DURATION	b
08	5 BC	70+ days	- 25°
09	AD 61	70 days	+ 70
10	64	75 days	+ 55
11	70	48 days	+ 45
17	185	20 months	0
19	247	156 days	+ 40
23	369	5 months	-
24	386	3 months	- 10
25	393	8 months	0
26	396	50+ days	- 25
27	402	2 months	+ 60
39	837 (B)	75 days	+ 65
45	1006	several years	+ 15
48	1054	22 months	- 5
58	1181	185 days	0
70	1572	16 months	0
72	1592 (A)	15 months	- 70
73	1592 (B)	3 months	0
74	1592 (C)	4 months	0
75	1604	12 months	+ 5

Table 3.2. Temporary stars of medium and long duration.

comets since there is a possible allusion to motion or a tail. The remaining high latitude stars are those of AD 70, AD 396, AD 837(B), and AD 1592(A). On the basis of high galactic latitude alone, it is tempting to classify these events as novae; there are certainly no SNRs at such high latitudes which could possibly correspond (see Chapter 4). In any case, even if they were supernovae, on account of their high galactic latitude they would have to be relatively nearby. Extreme brightness would thus be expected, but in no case is this reported. There is the possibility that the new star of AD 1592(A) was Mira Ceti. Two maxima are seemingly recorded for the star, about a year apart. Mira has a period of roughly 11 months, but the recorded position of the star is more than 10 degrees away. A second possible nova with a deep minimum in the light curve followed by a secondary maximum (similar to the recent nova DQ Her. of 1934) is the new star of AD 396. This was originally seen for a period of more than fifty days and was then recovered after a lapse of three months. However, as mentioned above, records for this period tend to be very brief so that the true nature of the object must be considered doubtful. Indeed, in the absence of additional information, the new stars of AD 70, 396, 837(B), and 1592(A) can, at best, be assigned as probable novae.

In looking for possible supernovae, we are thus left with the new stars of AD 185, 369, 386, 393, 1006, 1054, 1181, 1572, 1592(B), 1592(C), and 1604. These will be considered further in the latter half of the next chapter.

Chapter 4

THE SEARCH FOR THE REMNANTS OF SUPERNOVAE

In Chapter 1, the four principal observable remnants of a supernova were given as a pulsar, an expanding optical nebulosity, a region of radio emission, and an extended X-ray source. Of the more than 100 objects within our own galaxy positively identified as the remnants of supernovae, only the Crab Nebula and Vela remnants are known definitely to display all four features. For the remainder, the extended radio source is the most obvious remnant, and appears to be common to all events; because of obscuration effects, optical and X-ray emission may be detected only for the remnants of supernovae which occurred relatively nearby. The possible detection of a pulsar is subject to observational limitations, and in addition it is by no means yet certain that all supernova events produce an observable pulsar.

The radio remnants of supernovae came under investigation during the very earliest days of radio astronomy after World War II, and in fact the very first discrete stellar object (other than the Sun) identified as a radio source was the Crab Nebula, the remnant of the supernova of AD 1054. This identification was made by Bolton (1948) using a 'sea interferometer' mounted on a cliff over-looking the ocean near Sydney, Australia. That SNRs are one class of galactic radio source was established almost from the beginning of the science of radio astronomy, as was the nature of the radio emission from such remnants.

The discrete celestial radio sources emit a smooth 'continuum' of radio frequency energy (i.e. the emission intensity varies smoothly, with emission frequency). In addition, some sources show enhanced 'line' emission at certain characteristic frequencies. Continuum radio emission from our galaxy is known to come from two principal types of extended object.

A. Ionized Hydrogen Clouds. The interstellar hydrogen in the galaxy tends to be distributed in clouds, and this normally remains in its neutral (H I) state. However if a hot stellar source lies near or within such a cloud, ultra-violet radiation from the source tends to ionize the cloud, separating the hydrogen atoms into their constituent protons and electrons. When an electron and proton in random thermal motion approach each other at speed, they interact. The interaction is termed "free-free", since the particles are unbound before and after the interaction. The proton shows only a small deflection from its original path, but the lighter electron completely changes its direction emitting radiation known as thermal radiation, the wavelength of which depends upon the particle velocities and the distance at which they interact. In a hot gas the intensity of

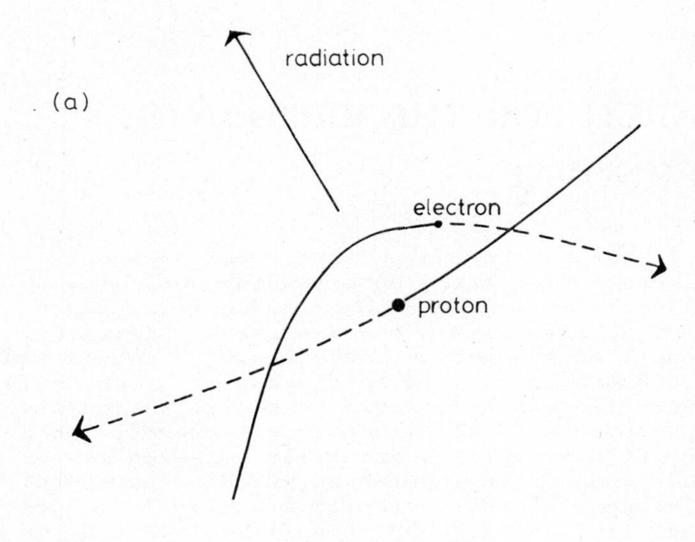

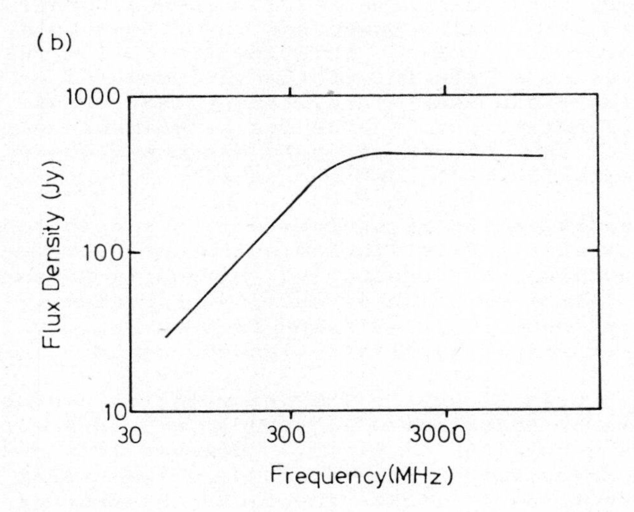

Fig. 4.1. (a) Thermal radiation is produced by the 'free-free' interaction of an electron and a proton. The proton shows only a small deflection from its original path, but the lighter electron completely changes its direction emitting radiation. (b) The radio-spectrum of the Orion Nebula, a typical ionized hydrogen region.

the thermal radiation remains essentially constant with wavelength, although beyond a certain point the intensity decreases with increasing wavelength (see Fig. 4.1). This typical 'thermal spectrum' is unique to radio sources associated with ionized hydrogen (H II) clouds within our galaxy. An additional feature of some such clouds is the line emission which may occur from ionized hydrogen regions during a transient recombination of an ion and electron (for example, the H109α recombination line at a frequency of approximately 5,009 MHz).

B. Supernova Remnants. The earliest observations of the Crab Nebula radio source showed that the intensity of the emitted energy increased with increasing wavelength. This is referred to as non-thermal radiation to distinguish it from the thermal radiation already described. Radio sources later identified as the remnants of Tycho's (AD 1572) and Kepler's (AD 1604) supernovae were also found to emit non-thermal radiation. The mechanism for producing such a spectral distribution remained, for many years, one of the major mysteries of radio astronomy. An explanation was eventually found by Shklovsky (1953), who suggested that the emission was produced by the so-called synchrotron mechanism. The synchrotron is a type of particle accelerator. When in a synchrotron electrons are accelerated to very high speed in a magnetic field, they radiate electromagnetic energy at a variety of wavelengths. The mechanism is the same in supernovae, where high energy electrons spiralling along magnetic field lines radiate light, while those of low energy radiate at radio wavelengths (see Fig. 4.2). The variation of emission intensity I with frequency ν follows a simple power law of the form $I(\nu) \sim \nu^{\alpha}$, with negative spectral index α. Apart from the unique spectral distribution, which is related to the energy distribution of the electrons which produce the emission, an additional characteristic of synchrotron radiation is that it displays a high degree of polarization, a property verified for the emission from SNRs (see Fig. 4.3).

Most extended non-thermal radio sources close to the galactic plane have morphological and spectral characteristics similar to the unambiguously identified galactic SNRs, and are consequently themselves possible candidates. However the recognition of a non-thermal spectrum is not sufficient grounds for classifying a radio source as a galactic SNR, since extragalactic radio sources also display this characteristic. Identification with optical filamentary structure and/or a historically recorded supernova may be possible for relatively nearby remnants. In the absence of such an identification, the final evidence sought is the spatial distribution of the radio brightness of the source. From a study of many radio remnants of supernovae observed with both good sensitivity and high resolution, it is found that their brightness can be modelled quite well by a uniform disc of emission with a superimposed annulus of 'limb-brightening'. Such a 'ring' source is indicative of a shell emitting region centred on the site of the original supernova outburst. Fig. 4.4 shows two examples of SNRs displaying this characteristic peripheral emission. It must be emphasised that these examples

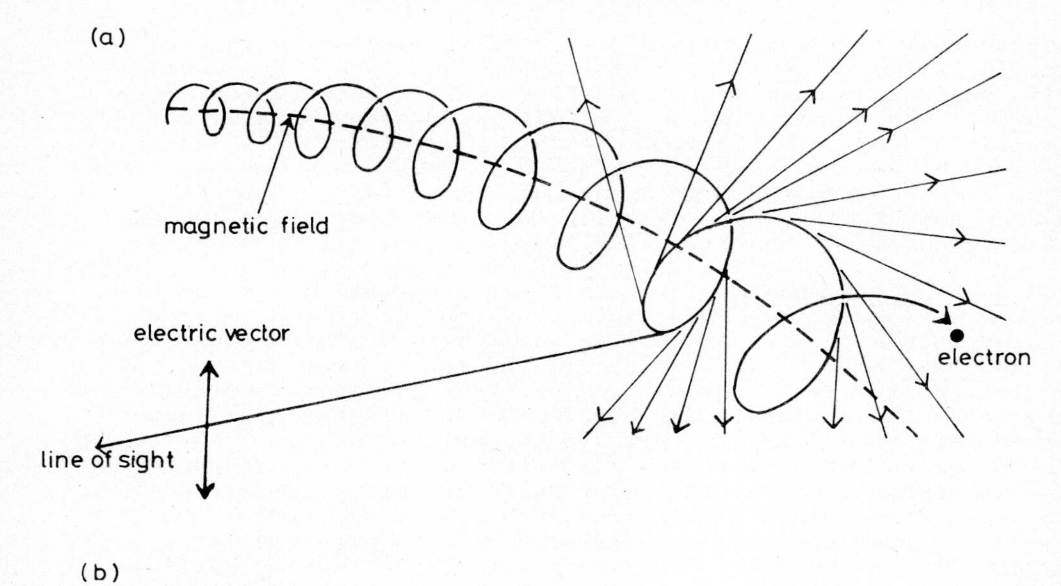

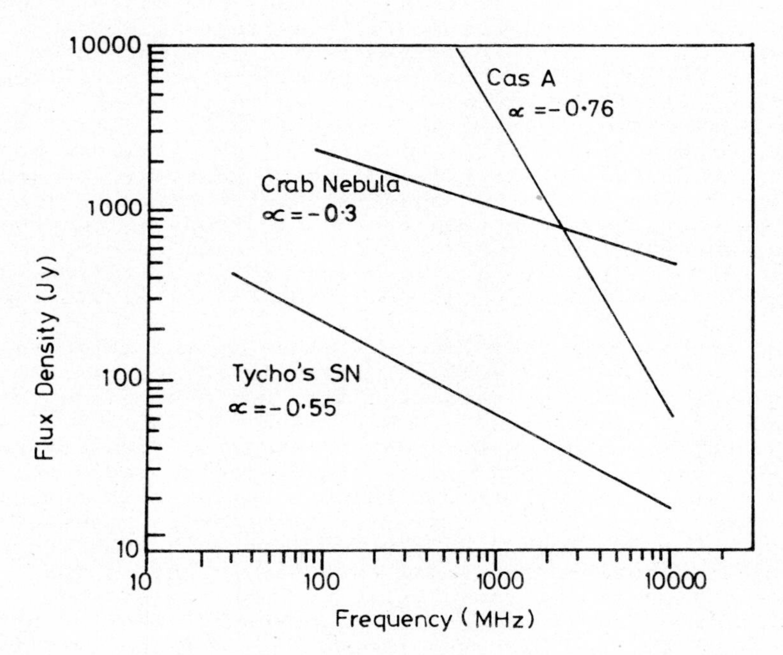

Fig. 4.2. (a) A fast electron in a magnetic field is constrained to move in a helical path wrapped around the field lines, and emits non-thermal radiation in the direction of motion. (b) Representative continuum radio spectra of three supernova remnants.

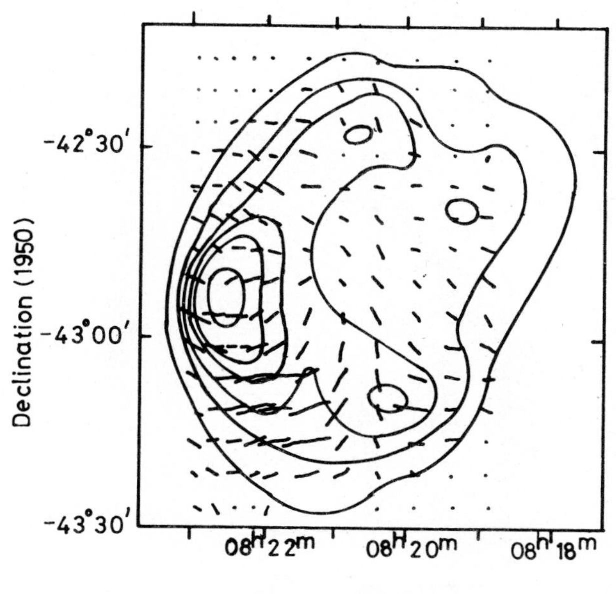

Fig. 4.3. A 2700 MHz map of the radiation from the supernova remnant Puppis A, showing vectors of polarization intensity. (From Milne 1971).

have been carefully chosen; many SNRs show only a part-ring or broken-shell emission, while a few (notably the Crab Nebula radio source) are amorphous displaying no limb-brightening, (more later).

To briefly summarize, in attempting to classify a celestial radio source as the remnant of a supernova one seeks (a) where possible, an identification with an optical filamentary structure and/or an historically recorded supernova, (b) a non-thermal spectrum and some degree of polarization indicative of the synchrotron process, and as a consequence of this property the absence of line emission which characteristically accompanies free-free continuum radio emission, (c) where resolution permits, the recognition of a 'ring' or 'part-ring' radio-brightness distribution, and (d) proximity to the galactic plane. On the basis of these classification criteria, the first comprehensive catalogues of galactic SNRs were produced independently by Milne (1970) and Downes (1971). A few of the sources in these catalogues have now been reclassified as ionized hydrogen clouds from improved observational data, and many new remnants have been identified by Clark, Caswell, and Green (1973, 1975) in a survey of the southern portion of the galactic plane made with the Mills Cross and Parkes

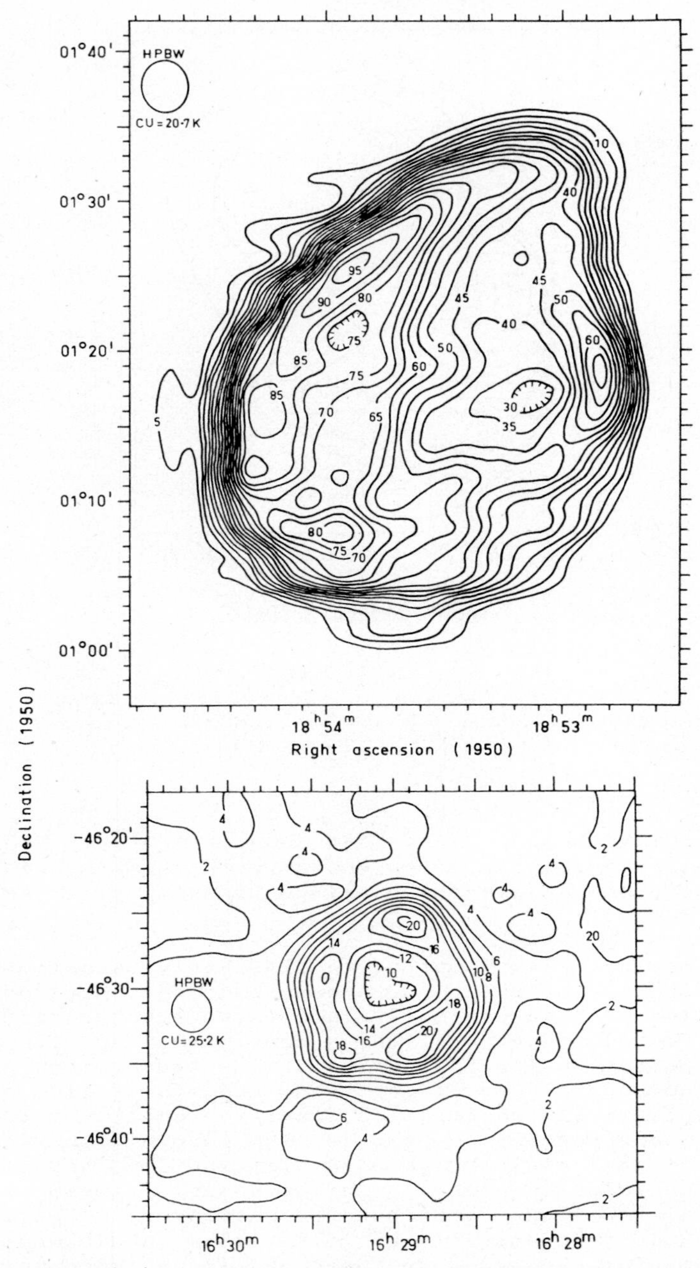

Fig. 4.4. Two examples of radio remnants displaying characteristic peripheral radio emission. (a) The SNR W44. (b) The SNR Kes 40. Both maps, at 408 MHz, were made with the Mills Cross radiotelescope of the University of Sydney.

radio telescopes in Australia . The most recent catalogues (see for example Green 1974, Clark and Caswell, 1976) now list about 120 galactic radio sources which are believed to be SNRs, and a galactic SNR catalogue is reproduced here as Table 4.1.

The table is composed as follows: column 1 gives the galactic source number of the SNR (galactic longitude followed by galactic latitude), and column 2 other catalogue numbers or common names; columns 3 and 4 give the right ascension and declination (epoch 1950.0) of the centroid of radio emission; column 5 gives the angular diameter of the source measured from the highest-resolution radio map available and referenced in column 6. Column 7 gives the spectral index α. The radio brightness integrated over the source yields the total flux density, and this is included in Column 8. (1 jansky (Jy) = 10^{-26} Wm^{-2} Hz^{-1}).

Woltjer (1970, 1972) has suggested that the dynamical evolution of an SNR shell may conveniently be divided into four phases, and this division has been adopted in most subsequent investigations.

Phase 1 (the free-expansion phase): The expansion of the supernova ejecta is essentially free following the supernova outburst (possibly with continuing injection of energy from a pulsar), and the appearance of the remnant depends on the initial conditions. A shock front forms at the leading edge of the swept-up interstellar gas but the mass of this gas is insufficient to decelerate the ejected matter noticeably.

Phase 2 (the adiabatic-expansion phase): The remnant is now dominated by swept-up matter, but radiative losses are still negligible and hence the total energy (kinetic plus thermal) is conserved. Shklovsky (1962) suggested that the expansion now resembles that of an adiabatic blast wave created by releasing energy at a point in a homogeneous gas. This phase continues only up to a time when radiation losses become important; thereafter the assumption of energy conservation is no longer valid.

Phase 3 (the isothermal-expansion phase): In this phase the structure of the object changes, as radiative cooling losses become significant. The matter that passes through the shock front cools rapidly and the density becomes high, with the formation of thin sheets or filaments which radiate in the optical and ultra-violet ranges. The remnant is now a rather thin shell ploughing through the interstellar medium. The thermal energy is small, and the shell may be considered to be moving with constant linear momentum.

Phase 4 (the extinction phase): When the expansion velocity of the shell becomes comparable to the thermal or random motion in the surrounding interstellar gas, the SNR gradually loses its identity and merges with the interstellar medium. It is of interest only in so far as it contributes to the intensity of the galactic background emission, and transfers kinetic energy to the interstellar medium.

Table 4.1

A Catalogue of 120 Galactic SNRs

(1) Galactic Source Number	(2) Catalogue number or name	(3) Right ascension (1950.0) h m s	(4) Dec. °	(5) Angular diameter (arc. min)	(6) Ref.	(7) α	(8) 1000 MHz Flux density (Jy)
G180.0-1.7	S147	05 36 00	27 30	171	16	-0.5	64
G184.6-5.8	Crab Nebula	05 31 31	21 59	5.2	33	-0.3	996
G189.1+2.9	IC 443	06 14 00	22 35	40	19	-0.4	161
G193.3-1.5	PKS 0607+17	06 05 50	16 40	80	7		27
G205.5+0.2	Monoceros	06 35 00	06 30	253	7	-0.5	115
G206.9+2.3	PKS 0646+06	06 46 00	06 30	80	24		5
G260.4-3.4	Puppis A	08 21 00	-42 50	47	14	-0.48	129
G261.9+5.5	PKS 0902-38	09 02 22	-38 29	40	18		8
G263.9-3.3	Vela X, Y, Z	08 32 30	-45 35	256	7	-0.5	1469
G287.8-0.5		10 45 00	-59 23	<42	20		
G290.1-0.8	MSH 11-61A	11 00 52	-60 37	12.6	27	-0.55	68
G291.0-0.1	MSH 11-62	11 09 49	-60 22	10.0	27	-0.35	16
G292.0+1.8	MSH 11-54	11 22 22	-58 59	5.4	27	-0.41	15
G293.8+0.6		11 33 05	-60 36	9.0	5	-0.58	5
G296.1-0.7		11 48 15	-62 27	16.0	5	-0.7	4
G296.5+10.0	PKS 1209-52	12 07 00	-52 07	81	31	-0.45	57
G296.8-0.3		11 55 48	-62 18	14.9	4	-0.62	9
G298.5-0.3		12 09 58	-62 36	3.7	27	-0.36	5
G298.6+0.0		12 11 18	-62 18	8.3	27	-0.30	4
G299.0+0.2		12 15 05	-62 08	10.5	5	-0.39	9

(1)	(2)	(3)	(4)	(5)	(6)	(7)	(8)
G302.3+0.7		12 42 54	-61 51	16.5	5	-0.36	5
G304.6+0.1	Kes 17	13 02 35	-62 26	6.9	27	-0.48	14
G308.7+0.0		13 38 05	-62 01	7.3	5	-0.35	12
G309.2-0.6		13 43 00	-62 36	12.6	5	-0.37	7
G309.8+0.0		13 47 03	-61 50	19.2	5	-0.51	17
G311.5-0.0		14 01 58	-61 43	3.9	27	-0.48	4
G315.4-0.3		14 32 00	-60 22	16.0	5	-0.47	10
G315.4-2.3	RCW 86	14 39 00	-62 17	39	4	-0.62	49
G316.3-0.0	MSH 14-57	14 37 43	-59 47	17.1	27	-0.32	28
G320.4-1.2	MSH 15-52	15 10 30	-59 00	25.8	27	-0.34	69
G321.9-0.3		15 16 45	-57 23	24.2	5	-0.34	13
G322.3-1.2	Kes 24	15 23 05	-57 56	5.8	4	-0.90	6
G323.5+0.1		15 25 05	-56 12	10.8	5	-0.41	3
G326.3-1.8	MSH 15-56	15 48 50	-56 00	36	6	-0.24	145
G327.1-1.1		15 50 35	-54 58	14.2	5	-0.36	8
G327.4+0.4	Kes 27	15 44 54	-53 39	21.0	4	-0.61	34
G327.6+14.5	PKS 1459-41	14 59 30	-41 45	34	23	-0.57	19
G328.0+0.3		15 49 33	-53 19	6.4	27	-0.55	3
G328.4+0.2	MSH 15-57	15 51 45	-53 08	4.0	27	-0.24	16
G330.0+15.0	Lupus Loop	15 09 00	-39 00	368	23		284
G330.2+1.0		15 57 20	-51 25	8.3	5	-0.30	7
G332.0+0.2		16 09 23	-50 49	12.0	27	-0.44	10
G332.4+0.1	MSH 16-51	16 11 38	-50 32	13.2	27	-0.51	25
G332.4-0.4	RCW 103	16 13 48	-50 54	9.4	27	-0.55	28
G335.2+0.1		16 23 50	-48 36	18.6	5	-0.46	18

Table 4.1. Continued

(1)	(2)	(3)	(4)	(5)	(6)	(7)	(8)
G336.7+0.5		16 28 30	-47 13	9.9	27	-0.37	7
G337.0-0.1	CTB 33	16 32 08	-47 30	7.6	27	-0.47	17
G337.2-0.7		16 35 45	-47 44	3.9	5	-0.67	2
G337.3+1.0	Kes 40	16 29 05	-46 29	11.8	4	-0.49	16
G337.8-0.1	Kes 41	16 35 15	-46 53	10.6	27	-0.51	16
G338.2+0.4		16 34 40	-46 16	11.7	27	-0.42	2
G338.3-0.1		16 37 25	-46 27	8.2	27	-0.66	7
G338.5+0.1		16 37 20	-46 12	12.4	27	-0.33	27
G339.2-0.4		16 41 50	-46 03	9.9	5	-0.20	6
G340.4+0.4		16 43 00	-44 34	6.4	5	-0.41	6
G340.6+0.3		16 44 05	-44 30	4.9	5	-0.36	5
G341.9-0.3	MSH 16-4_8_	16 51 20	-43 54	6.1	3	-0.59	4
G344.7-0.1		17 00 15	-41 37	7.8	5	-0.51	3
G346.6-0.2		17 06 45	-40 06	8.0	5	-0.49	10
G348.5+0.1	CTB 37A	17 11 12	-38 26	8.0	6	-0.33	72
G348.7+0.3	CTB 37B	17 10 45	-38 06	5.1	6	-0.30	26
G349.7+0.2		17 14 37	-37 23	1.7	28	-0.49	20
G350.0-1.8		17 23 45	-38 20	28.9	5	-0.51	31
G350.1-0.3		17 17 40	-37 24	4.1	5	-0.73	6
G351.2+0.1		17 19 00	-36 09	6.2	5	-0.38	6
G352.7-0.1		17 24 20	-35 05	6.4	5	-0.57	6
G355.9-2.5		17 42 30	-33 43	11.2	5	-0.51	8
G357.7-0.1	MSH 17-3_9_	17 37 04	-30 56	5.2	28	-0.43	37
G4.5 +6.8	Kepler's SN	17 27 41	-21 26	3.2	15	-0.58	20
G5.3 -1.0	A4	17 58 30	-24 50	15.0	6	-0.2	32

(1)	(2)	(3)	(4)	(5)	(6)	(7)	(8)
G 6.4-0.1	W28	17 57 35	-23 27	49	27	-0.38	327
G 7.7-3.7		18 14 15	-24 05	19.5	Unpublished	-0.25	10
G10.0-0.3		18 05 42	-20 25	6.0	27	-0.34	2
G11.2-0.3		18 08 31	-19 26	4.2	28	-0.56	22
G11.4-0.1		18 07 50	-19 06	7.0	5	-0.48	6
G12.0-0.1		18 09 16	-18 38	5.4	5	-0.71	3
G15.9+0.2		18 16 07	-15 02	5.0	5	-0.56	5
G18.8+0.3	Kes 67	18 21 00	-12 25	15.0	6	-0.36	27
G21.8-0.6	Kes 69	18 30 05	-10 09	22.8	27	-0.54	68
G22.7-0.2		18 30 35	-09 13	24.6	27	-0.57	32
G23.3-0.3	W41	18 31 40	-08 51	21.0	27	-0.50	59
G23.6+0.3		18 30 25	-08 14	7.2	27		5
G24.7+0.6		18 31 30	-07 08	14.0	27	-0.59	16
G24.7-0.6		18 36 00	-07 37	14.6	5	-0.49	8
G27.4+0.0	Kes 73	18 38 36	-04 59	4.4	3	-0.45	3
G29.7-0.2	Kes 75	18 43 48	-03 02	2.4	27	-0.71	10
G31.9+0.0	3C 391	18 46 47	-00 59	4.8	28	-0.57	51
G32.0-4.9	3C 396.1	19 04 30	-03 06		2	-0.5	
G32.8-0.1	Kes 78	18 48 55	-00 13	17.2	4	-0.20	11
G33.6+0.1	4C 00.70	18 50 05	00 37	9.2	4	-0.60	21
G34.6-0.5	W44	18 53 35	01 17	27.2	6	-0.28	232
G39.2-0.3	3C 396	19 01 35	05 22	6.6	28	-0.49	19
G39.7-2.0	W50	19 09 15	05 12	›30	6		›25
G41.1-0.3	3C 397	19 05 08	07 04	3.6	4	-0.49	19
G41.9-4.1	PKS 1920+06	19 20 00	06 00	164	24	-0.5	19

(1)	(2)	(3)	(4)	(5)	(6)	(7)	(8)
G43.3 -0.2	W49B	19 08 43	09 01	4.2	28	-0.47	35
G46.8 -0.3		19 15 45	12 05	15.6	4	-0.42	14
G47.6 +6.1	CTB 63	18 54 00	15 45	83.4	24	-0.5	
G49.2 -0.5	W51	19 21 30	13 57	26.6	26	-0.25	160
G53.7 -2.2	3C 400.2	19 36 15	17 07	26.6	6	-0.32	9
G54.4 -0.3	HC 40	19 31 00	18 55	43	30	-0.49	27
G55.7 +3.4		19 19 40	21 37	17.4	12	-0.5	1
G65.7 +1.2	DA 495	19 50 05	29 17	18.6	32	-0.64	6
G74.0 -8.6	Cygnus Loop	20 49 00	30 30	180	21	-0.45	211
G74.9 +1.2	CTB 87	20 14 05	37 04	6.5	11	-0.47	9
G78.1 +1.8	DR 4	20 20 45	40 04	12.2	17	-0.65	55
G82.2 +5.4	W63	20 17 15	45 25	77	30	-0.7	133
G89.0 +4.7	HB 21	20 43 20	50 29	100	22	-0.40	259
G93.2 +6.7	4C(T) 55.38.1	20 51 00	55 07	25.0	Unpublished	-0.37	8
G93.6 -0.2	CTB 104A	21 26 50	50 33	61	30	-0.35	37
G94.0 +1.0	3C 434.1	21 23 30	51 40	26.6	30	-0.53	12
G111.7-2.1	Cas A	23 21 11	58 33	4.2	25	-0.76	3390
G117.3+0.1	CTB 1	23 56 40	62 09	30	32	-0.58	8
G119.5+10.0	CTA 1	00 02 35	72 20	130	1	-0.2	41
G120.1+1.4	Tycho's SN	00 22 37	63 52	7.9	29	-0.55	56
G130.7+3.1	3C 58	02 01 53	64 35	5.2	13	-0.1	35
G132.4+2.2	HB 3	02 15 00	62 25	72	30	-0.52	62
G160.5+2.8	HB 9	04 57 00	46 36	139	8	-0.44	135
G166.0+4.3	VRO 42.05.01	05 22 50	42 52	44	9	-0.40	7
G166.2+2.5	OA 184	05 15 30	41 47	76	10	-0.55	10

References for Table 4.1.

1. Caswell (1967)
2. Caswell (1970a)
3. Caswell and Clark (1975)
4. Caswell, Clark and Crawford (1975)
5. Clark, Caswell and Green (1973, 1975)
6. Clark, Green and Caswell (1975)
7. Day, Caswell and Cooke (1972)
8. Dickel and McKinley (1969)
9. Dickel, McGuire and Yang (1965)
10. Dickel and Yang (1965)
11. Duin et al. (1975)
12. Goss and Schwarz (1971)
13. Goss, Schwarz and Wesselius (1973)
14. Green (1971)
15. Gull (1975)
16. Haslam and Salter (1971)
17. Higgs and Halperin (1968)
18. Hill (1967)
19. Hill (1972)
20. Jones (1973)
21. Keen et al. (1973)
22. Kundu (1971)
23. Milne (1971a)
24. Milne and Hill (1969)
25. Rosenberg (1970)
26. Shaver (1969)
27. Shaver and Goss (1970a)
28. Slee and Dulk (1974)
29. Strom and Duin (1973)
30. Velusamy and Kundu (1974)
31. Whiteoak and Gardner (1968)
32. Willis (1973)
33. Wilson (1970)

Analytical solutions are available for the various phases. Rosenberg and Scheuer (1973), Chevalier (1974), and others, have attempted numerical computation of the full dynamical development; this work has emphasized the uncertainties of assigning observed SNRs to one of the above phases, since the transitions between them are complex and with unknown time scales. Nevertheless, division into the four phases provides a useful first approximation for a study of SNR evolution.

Although the spectrum and polarization of the radio emission from SNRs confirm that it is synchrotron radiation, there is no single completely satisfactory theory yet for the origin and evolution of the particles and fields. There is increasing evidence that these might be dominated by quite distinct physical processes at various stages of a remnant's evolution.

Gull (1973a) has suggested that during Phase 1, instabilities at the interface between the ejecta and the interstellar medium result in a region of enhanced magnetic field which gives a shell source of non-thermal radiation as relativistic particles diffuse into the interstellar medium; during this phase the radio emission is expected to rise to a maximum and subsequently decay. This model is found to describe well the radio properties of such young SNRs as Cas A and the remnant of Tycho's supernova (Gull 1973b).

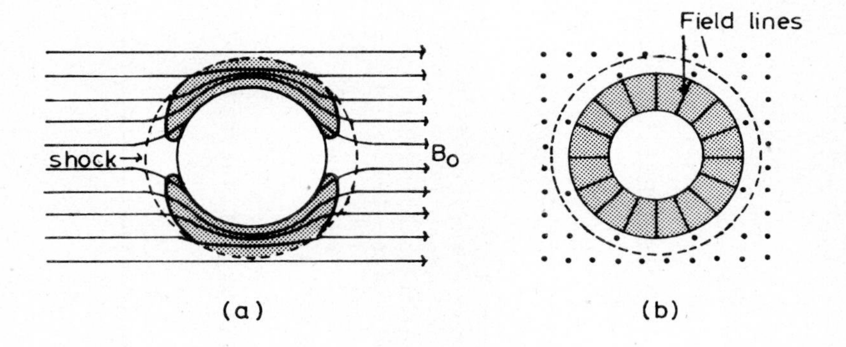

Fig. 4.5. The evolutionary model of van der Laan.
(a) The interstellar magnetic field B_0 perpendicular to the line of sight. The region of maximum radio emission, shown shaded, appears as two 'lobes' containing the compressed magnetic field.
(b) The interstellar magnetic field along the line of sight. The region of maximum emission is a ring containing the transverse component of the distorted magnetic field.
(From Whiteoak and Gardner, 1968).

For older remnants the interstellar magnetic field and relativistic cosmic ray electrons are expected to give rise to non-thermal radiation after compression at the shock front, as in the model of van der Laan (1962). The model is illustrated in Fig. 4.5. Fig. 4.5(a) shows the case where the interstellar magnetic field is perpendicular to the line of sight. The maximum radio emission then results from the two 'lobes' containing the compressed magnetic field. Many SNRs show this characteristic double-lobe structure, and an example is shown as Fig. 4.6. Measurements of polarization direction (always perpendicular to the field orientation) would then be expected to be radial. Fig. 4.5(b) shows the case where the interstellar magnetic field is exactly along the line of sight. Here synchrotron emission is observed only where the distorted field has a transverse (i.e. apparently radial) component, so that the radio emission would be a complete 'ring' with zero intensity at the centre and the direction of polarization circumferential.

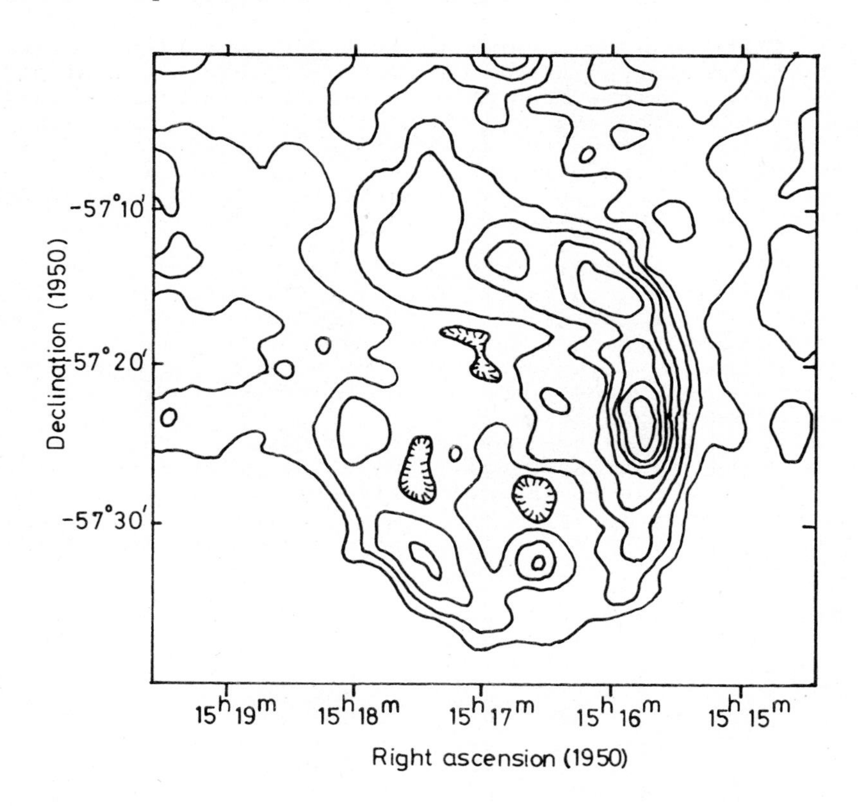

Fig. 4.6. A 408 MHz map of the remnant G321.9-0.3, displaying the characteristic double-lobe structure of an SNR viewed across the interstellar magnetic field.

The distribution of SNRs shows a concentration close to the galactic plane, and the majority lie within little more than the solar distance from the galactic centre. A source count from the catalogue of SNRs gives the following distribution in galactic latitude b: 75 sources for $|b| < 1^{\circ}$; 34 sources for $1^{\circ} \leqslant |b| < 5^{\circ}$; 7 sources for $5^{\circ} \leqslant |b| < 10^{\circ}$; and only 4 sources for $|b| \geqslant 10^{\circ}$. The highest known value of b for a young galactic SNR is 15°. Although most galactic SNR surveys have been restricted to within a few degrees of the plane, the concentration of remnants towards the plane is not believed to be a consequence of this; indeed, Henning and Wendker (1975) have concluded from a search of available high-latitude radio data that there are no additional high-latitude sources which would definitely be assigned as SNRs. The existence of a few SNRs at large distance, say >300 pc, from the plane (for example, the remnant of the supernova of AD 1006 - more later) negates the suggestion that only supernovae occurring near the galactic plane leave observable remnants.

The first celestial X-ray source was discovered in 1962. The present X-ray catalogues contain over 200 objects, several of which are SNRs. X-rays from SNRs result from thermal radiation and line emission from the heated interstellar material swept up by the expanding blast wave; in addition, synchrotron radiation is recognised as the principal source of X-ray emission in the Crab Nebula, although this X-ray emission mechanism has not been confirmed for any other SNR.

Following the free-expansion phase, the leading edge of the expanding shell of ejecta experiences significant deceleration with resulting increased density. The faster material following enters this denser region from behind, with resultant heating of the gas and forming a second shock front behind the outer dense shell. In the reference frame of the expanding shell this second shock front appears to be moving inward; hence it is usually referred to as the 'reverse shock wave'. McKee (1974) has suggested that the heated gas behind the reverse shock wave may be the source of soft thermal X-ray emission in young SNRs, while harder thermal X-ray emission might be detected behind the leading edge shock wave where the gas is heated to temperatures in the range 10^{7} - 10^{9} K.

During the adiabatic phase of a remnant's expansion, the velocity of the shock wave falls as the mass of swept up interstellar matter increases. The temperature of gas behind the leading edge shock wave falls to about 10^{6} - 10^{7} K, and the X-ray emission is shifted to longer wavelengths with increased intensity due to the larger mass of radiating hot gas. Concentration of the hot gas in an expanding shell results in peripheral brightening of the source (see Fig. 4.7), while inhomogeneities in the interstellar medium are usually invoked to explain the 'patchiness' of the X-ray emission with bright regions resulting from the encounter of the expanding shock wave with local interstellar density enhancements.

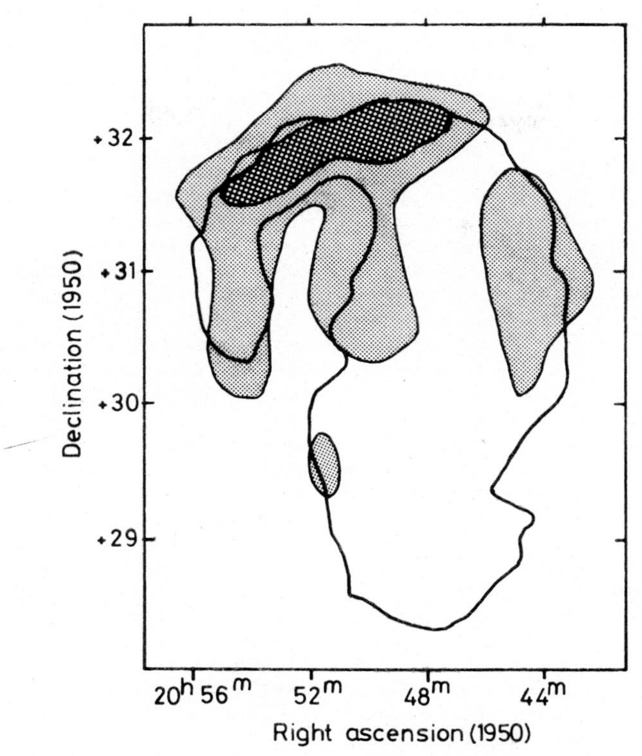

Fig. 4.7. The Cygnus Loop, showing the outline of the region of continuum radio emission with the region of X-ray emission shaded ; (a simplified version of a diagram from Seward et al, 1976).

At the present time eight nearby SNRs have been definitely confirmed to be X-ray emitters. X-ray emission for the Crab Nebula is recognised to be synchrotron in nature, with characteristic non-thermal spectrum and significant polarization. At X-ray wavelengths, the young SNRs of Tycho's supernova and Cas A both exhibit thermal spectra which may be explained by the 'reverse shock wave' model. The remnant of the supernova of AD 1006 shows weak thermal emission, while the older remnants Vela, Puppis A, IC443, and the Cygnus Loop all display more intense thermal X-ray emission in accordance with predictions from the adiabatic expansion model. For recent reviews of the X-ray remnants of supernovae, see Charles and Culhane (1975), and Clark and Culhane (1976).

While the early chapters of the history of X-ray astronomy have been dominated by the spectacular discoveries relating to the existence of neutron stars and black holes in binary systems, it is our belief that the later chapters will reveal great advances in astrophysics achieved through the observation of the X-ray

SNRs. The X-ray remnants of supernovae appear to be amongst the most valuable of astronomical antiquities, capable of revealing information on the initial supernova outburst, the evolution of the remnants, and the structure and composition of the interstellar medium.

There are three main classes of optical emission nebulae in the galaxy, although the atomic processes that produce the optical emission for each of these three classes are basically the same. Ionized hydrogen clouds represent the first class. As mentioned earlier, the ultraviolet emission from a hot central star will ionize the cloud. Transient recombination of a proton and an electron can then take a variety of forms. An electron can be captured at any energy level of the atom, and then falls rapidly, via a variety of possible intermediate states, to the ground state. The transition from the hydrogen atom's third energy level to the second energy level produces the most intense visible radiation, at a wavelength of 6,563 angstroms - the so-called Hα emission. The second class of emission nebulae are the planetary nebulae. The shell of gas emitted by a nova may be ionized by emission from the post-nova. Recombination in the expanding shell is then evidenced by emission at wavelengths characteristic of the chemical elements in the shell. The third main class of emission nebulae are the optical remnants of supernovae, where the source of energy is not necessarily the photons from a central star. In the case of the Crab Nebula, ionization is produced by synchrotron radiation. However for the majority of optical remnants, the ionization is believed to result from the interaction of the expanding shock wave with inhomogeneities in the interstellar gas; the light we observe is then generated behind the shock wave, where the ions and electrons are cooling and recombining. The relative abundances of different elements in a remnant can then be interpreted in terms of the strengths of its various emission lines.

Of the 120 galactic supernova remnants, only 24 display any optical emission - the most recent catalogue of these optical remnants is that of van den Bergh, Marscher, and Terzian (1973). The optical remnants of supernovae, the most spectacular of which include S147 (see Plate 6), the Crab Nebula, IC443, Vela, and the Cygnus Loop, are amongst the most beautiful nebulosities in the skies.

As noted in Chapter 1, only supernovae which occurred on the near-side of the Galaxy could be expected to have been observed from the Earth ; the brightness of an outburst then gives a crude estimate of its distance. In looking for the probable remnants of the historical supernovae, one therefore requires distance estimates for the SNRs. Such estimates may be obtained in a variety of ways, but only the two most common techniques will be described briefly here.

The most reliable technique for SNR distance determination uses the absorption of the continuum radio emission from the SNR by the neutral interstellar hydrogen (H I) along the line-of-sight.

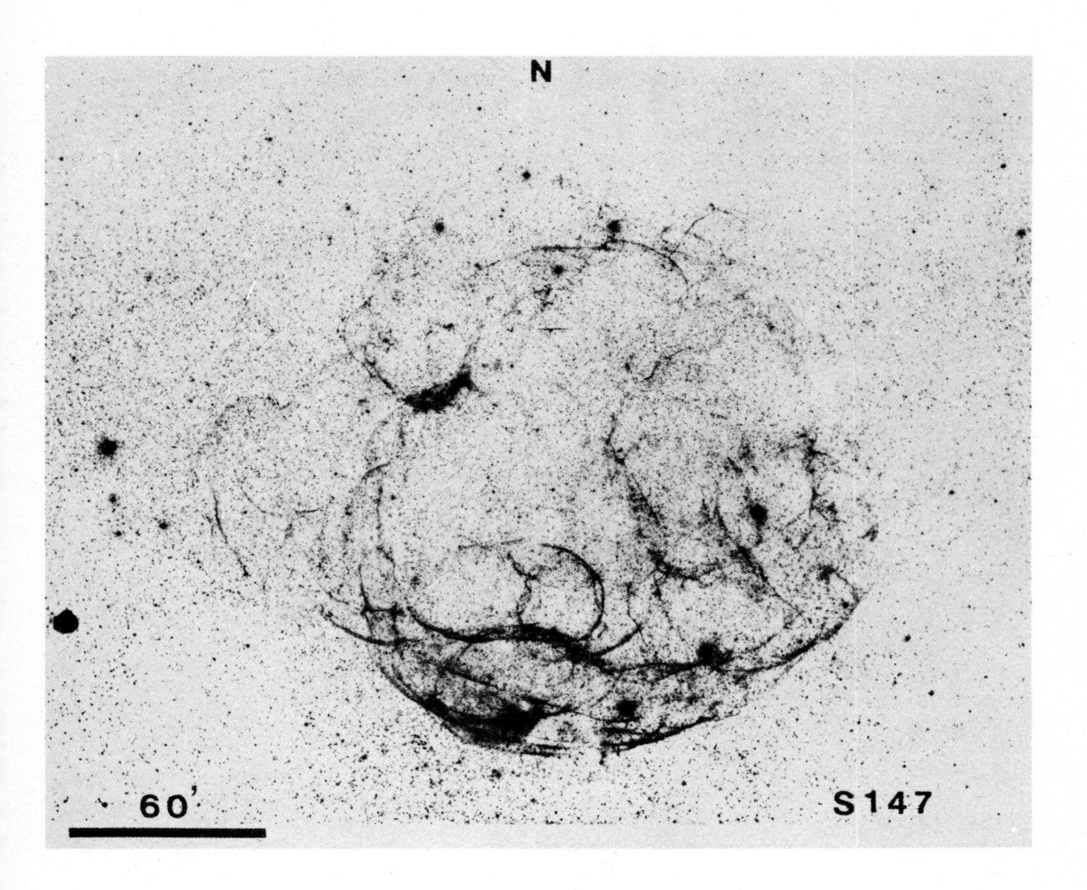

Plate 6. The optical remnant S147.

The 21 cm (1420 MHz) spectral line from neutral hydrogen was first detected in emission by Ewen and Purcell (1951), and in absorption by Hagen and McClain (1954). The Doppler shift of the line observed in various directions gives the velocity with which the neutral hydrogen is moving towards or away from us, and with a suitable frequency (velocity) - distance model radio astronomers have been able to derive from such observations a picture of the spiral structure of our Galaxy (see for example Kerr, 1968). For neutral-hydrogen measurements in the direction of certain SNRs, well-defined absorption minima are exhibited; these are interpreted as being due to neutral-hydrogen clouds in the spiral arms along the line-of-sight. The positions of the intervening spiral arms for a particular SNR may be determined from a reference profile slightly displaced from the direction of the SNR, and the absorption feature corresponding to the most distant cloud then gives the minimum distance to the SNR. This H I absorption technique has been successfully applied to give minimum distances to about 30 galactic SNRs (see for example Caswell et al, 1976).

In the second technique, a distance for an SNR may be inferred from comparing its radio surface brightness (Σ) and angular diameter with those for SNRs with well determined values. (The mean surface brightness is defined as the integrated flux divided by the angular area of the source, and is measured in units of $Wm^{-2}\ Hz^{-1}\ sr^{-1}$). It has long been recognised (e.g. Shklovsky 1960 a, b) that the mean surface brightness of an SNR at a particular radio frequency is an extremely valuable observational parameter because it is distance-independent, and to a first approximation an intrinsic property. After reaching a maximum shortly after the birth of the supernova, Σ may be expected to decrease monotonically with time. (It has been conjectured that in later phases, as different factors become important, an increase might occur; however there is no real observational evidence for this). The outer diameter D of the expanding SNR will <u>increase</u> monotonically with time. Using SNRs of known distance we may then plot their distribution in the Σ - D plane, when it is found that they appear to follow a common evolutionary track (see for example Clark and Caswell, 1976). The expected linear diameter of an SNR of unknown distance but with measured surface brightness can then be found from the Σ - D plot, and a distance thus estimated from the measured angular diameter of the source. The Σ - D approach is at present an empirical one with little theoretical foundation. The limits of the technique are discussed by Clark and Caswell (1976); despite these limitations Σ - D distance estimates are believed accurate to within $\pm$ 30% for the majority of SNRs.

The catalogue of radio remnants of Clark and Caswell (1976) has been searched to see which sources, being of small diameter and high surface brightness, might be young enough to have originated in supernovae that have occurred within, say, the last two thousand years and which therefore might have been historically recorded. The variation of remnant diameter with time has not been well understood in the past because of the shortage of reliable age calibrators, although Clark and Caswell (1976), on the basis of various assumptions detailed in the original reference, have derived the empirical relationship -

$$D \simeq 0.9\, t^{2/5}$$

from a statistical analysis of their catalogue, where D is the remnant diameter in parsec, and t is the time in years since the explosion. (This diameter - time relationship will be further investigated in Chapter 12). From the equation, an age of 2000 years corresponds to a remnant of diameter about 20 pc (with large uncertainty).

We have previously noted that a new star is most unlikely to be discovered with the unaided eye if it is fainter than about magnitude +3. As outlined in Chapter 1, a supernova near the plane will reach magnitude +3 only if it occurs at a distance of less than about 7 to 8 kpc. In our search for 'young' SNRs of supernovae which might have been recorded historically, we therefore limited ourselves to those with estimated distances less than 10 kpc, well beyond the expected limit.

The SNRs from the Clark and Caswell catalogue which meet the above mentioned criteria (distance $\leqslant$10 kpc, diameter $\leqslant$ 20 pc) are listed in Table 4.2. The Table is composed as follows : Column 1 gives the Galactic source number of the SNR, and a common name or alternative catalogue number; column 2 gives its declination (epoch 1950.0) - important in considering a possible historical detection; columns 3 and 4 give source distance and diameter estimates found by the technique given in column 5 - where such estimates are lower limits, the Σ - D parameters for the source are also given in parentheses; the references for these data are given in column 6. The sources are also shown plotted (and identified by Table number, or year of outburst for the well-established supernovae) on the simplified 'bird's-eye' view of the Galaxy in Fig. 4.8. SNRs with lower limits for distances are indicated by arrows directed away from the lower limit value.

The first source in Table 4.2. is the well known remnant of AD 1054 (to be discussed in detail in Chapter 8). Sources 2, 3, and 4 lie so far south as to reduce the likelihood of their detection from Europe, the Middle or Far East. Source 5 will be discussed in Chapter 5 as the probable remnant of the supernova of AD 185. Source 6 is at an extreme southern declination, and has a lower limit diameter which may imply an age $>$ 2000 years, Source 7 is the remnant of the supernova of AD 1006, and will be discussed in Chapter 7.

Table 4.2

Young SNRs within 10 kpc of the Sun

	(1) Galactic source number (Other name)	(2) Declination (1950.0)	(3) Distance (kpc)	(4) Mean linear diameter (pc)	(5) Notes	(6) References.
1.	G184.6-5.8 (Crab Nebula)	$+21^{\circ}59'$	2.2	3.0	A	7
2.	G290.1-0.8 (MSH 11-61A)	$-60^{\circ}39'$	$\geqslant$3.4 (5.8)	$\geqslant$12.5 (21.1)	B	2
3.	G292.0+1.8 (MSH 11-54)	$-58^{\circ}59'$	$\geqslant$3.7 (13.4)	$\geqslant$5.8 (21.0)	B	1
4.	G304.6+0.1 (Kes 17)	$-62^{\circ}26'$	9.7	19.5	B	1
5.	G315.4-2.3 (RCW 86)	$-60^{\circ}22'$	2	22	C	8
6.	G326.3-1.8 (MSH 15-56)	$-56^{\circ}02'$	$\geqslant$1.5 (3.2)	$\geqslant$15.7 (33.4)	B	1
7.	G327.6+14.5 (SN 1006)	$-41^{\circ}45'$	1.3	12.8	C	5
8.	G332.4-0.4 (RCW 103)	$-50^{\circ}58'$	3.3	9.0	B	1

9.	G348.5+0.1 (CTB 37A)	$-38^{\circ}26'$	$\geq$6.7 (7.1)	$\geq$15.6 (16.4)	B	
	G348.7+0.3 (CTB 37B)	$-38^{\circ}06'$	$\geq$6.7 (11.6)	$\geq$9.7 (17.2)	B	1
10.	G357.7-0.1 (MSH 17-39)	$-30^{\circ}56'$	$\geq$6 (9.8)	$\geq$9.1 (14.9)	B	6
11.	G4.5+6.8 (Kepler)	$-21^{\circ}26'$	10	9.3	C	4
12.	G11.2-0.3	$-19^{\circ}26'$	$\geq$5 (12.1)	$\geq$6.1 (14.8)	B	6
13.	G29.7-0.2 (Kes 75)	$-03^{\circ}02'$	$\geq$6.6 (17.9)	$\geq$4.6 (12.5)	B	1
14.	G33.6+0.1	$+00^{\circ}37'$	$\geq$7 (9.4)	$\geq$18.7 (25.1)	B	1
15.	G41.1-0.3 (3C397)	$+07^{\circ}04'$	$\geq$7.5 (13.7)	$\geq$7.9 (14.3)	B	6
16.	G111.7-2.1 (Cas A)	$+58^{\circ}33'$	3.3	3.7	B,C	4
17.	G120.1+1.4 (Tycho)	$+63^{\circ}52'$	$\geq$6 (7.2)	$\geq$13.8 (16.5)	B	3,9
18.	G130.7+3.1 (3C58)	$+64^{\circ}35'$	$\geq$8 (11.1)	$\geq$12.1 (16.8)	B	3,9

Notes : A : optical radial velocity and proper motion

B : H I absorption, kinematic distance

C : From optical and other considerations

References to Table 4.2

1. Caswell et al (1976
2. Goss et al (1972)
3. Goss et al (1973)
4. Ilovaisky and Lequeux (1972)
5. Minkowski (1966)
6. Radhakrishnan et al (1972)
7. Trimble (1968)
8. Westerlund (1969)
9. Williams (1973)

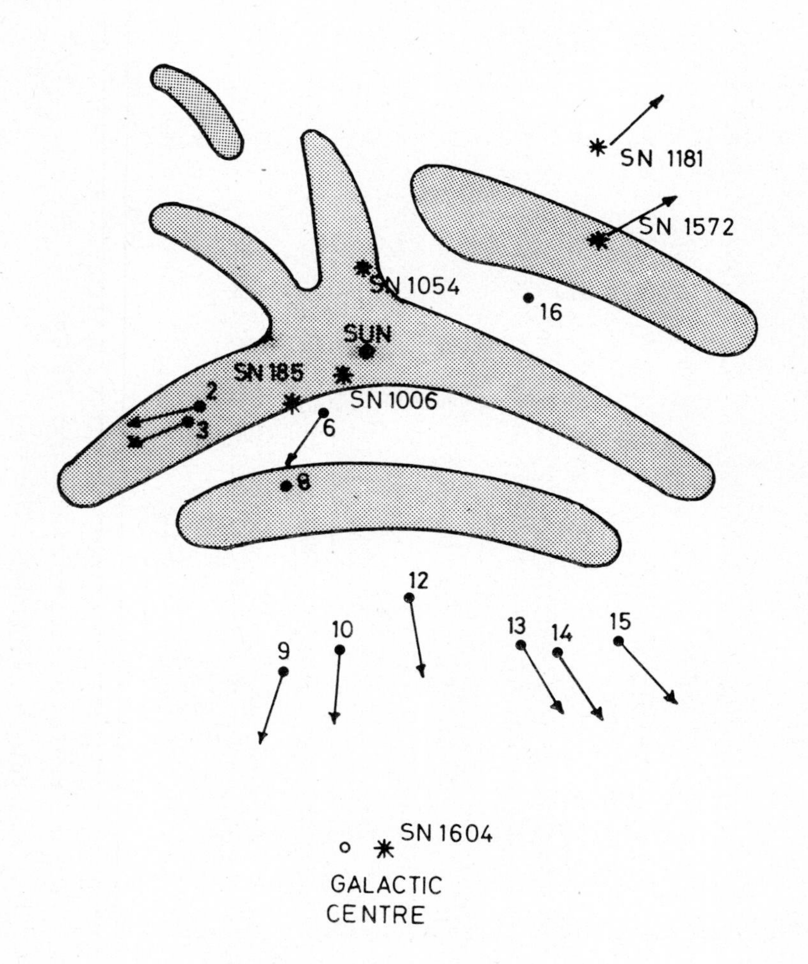

Fig. 4.8. A "bird's-eye view" of the Galaxy, showing the estimated positions of the SNRs from Table 4.2.

Source 8 has a well established diameter which would suggest it being a young SNR, and a distance at which a historical observation might have been expected although again the southerly declination might have reduced the likelihood of it being sighted from the northern hemisphere - this possibility is discussed in more detail later. Source 9 (in fact probably two adjacent SNRs) will be discussed in Chapter 6 as the possible remnant of the proposed supernova of AD 393. Source 10 is close to the galactic centre and lying in the plane, so that a visual detection would have been unlikely. Source 11 is the remnant of the supernova of AD 1604 (see Chapter 11). Sources 12, 13 and 15 all have only minimum H I absorption kinematic distance estimates, but large (>10 kpc) Σ - D distances making them poor candidates for detection. However in Chapter 6 we introduce the possibility that source 12 is the remnant of a supernova in AD 386. The Σ - D diameter for source 14 exceeds the 20 pc (equivalent to age 2000 years) limit discussed above. Source 16, Cas A, is almost certainly the remnant of a young, nearby but undetected, supernova. Source 17 is the remnant of the supernova of AD 1572 (see Chapter 10) and source 18 is probably the remnant of the supernova of AD 1181 (see Chapter 9). Thus in Table 4.2, apart from the proposed remnants of the long-duration new stars (all believed to be supernovae), only source 8 presents itself as a likely candidate for historical detection, with sources 12 through 15 less likely candidates. There is no source in the Table which could correspond to the event of AD 1069 to which we drew attention in Chapter 3.

Of the long duration new stars of interest listed in Chapter 3, only those of AD 369, and AD 1592 B and C remain to be accounted for. The position of the star of AD 369 is so poorly recorded that we cannot even deduce an approximate galactic latitude. The probable declination of this star was around $+65^{\circ}$, but the right ascension is in considerable doubt. However north of $+50^{\circ}$ there are no 'young' SNRs, other than G111.7-2.1 (Cas A), G120.1+1.4 (AD 1572), and G130.7+3.1 (AD 1181). We thus conclude that the star was probably a slow nova. Both stars of AD 1592 appeared in Cassiopeia, the former in the neighbourhood of α Cas and the latter close to β Cas. This area has been well surveyed for SNRs, and none are known apart from Cas A. Brosche (1967) and Chu Sun-il (1968) independently suggested that Cas A (G111.7-2.1) is the remnant of the star AD 1592C, but the preferred age of the remnant is much less than this (see for example Gull 1973b). It seems likely that both stars were slow novae.

For most of the time since AD 1280 the Chinese capital has been Peking ($\approx 40^{\circ}$ North). From this latitude a supernova south of declination 50° would not have been sighted. This is a possible explanation for the non-detection of the star which produced the young SNR G332.4-0.4, previously noted as a likely candidate for historical observation because of its proximity to the Earth. Alternatively the supernova may have occurred near the time of conjunction with the Sun, reducing the likelihood of its detection.

In the following chapters we present detailed discussions of the 8 historically-recorded new stars of long duration (i.e. of AD 185, 386, 393, 1006, 1054, 1181, 1572 and 1604) proposed, with varying degrees of certainty, as possible supernovae.

Chapter 5

THE GUEST STAR WITHIN THE SOUTHERN GATE

The new star of AD 185 is the earliest recorded for which there are any grounds for supposing it to be a supernova. The star was reported only in China, and all we know about it is contained in a single text.

No reference to the guest star is expected from Japan or Korea at such an early period. In the former country we are still in the legendary era. A number of astronomical records from Korean history around this time are accessible (in the Samguk Sagi), but these are of doubtful reliability; in any event, there is no mention of the star. As it was only visible south of about latitude 35°N (it appeared in the constellation of Centaurus), Europe can be ruled out as a possible source of observations. There seems a possibility that the star might have been carefully observed in Alexandria (31°), the home of the great astronomer Claudius Ptolemy half a century before, but, if so, no account of it has survived the ravages of time.

The solitary Chinese description of the guest star is to be found in part of the astronomical treatise (Chapter 22) of the Hou-han-shu. A translation of the text is as follows :- "2nd year of the Chung-p'ing reign period (of Emperor Hsiao-ling), 10th month, day kuei-hai, a guest star appeared within (chung) Nan-mên. It was as large as half a mat; it was multicoloured (lit. 'It showed the five colours') and it scintillated. It gradually became smaller and disappeared in the 6th month of the year after next (hou-nien). According to the standard prognostication this means insurrection. When we come to the 6th year, the governor of the metropolitan region Yüan-shou punished and eliminated the middle officials. Wu-kuang attacked and killed Ho-miao, the general of chariots and cavalry, and several thousand people were killed". The date of appearance of the guest star corresponds to AD 185 December 7.

Before commenting on the above record, it is appropriate to discuss its historical and astronomical background.

The various treatises of the Hou-han-shu were compiled by Szŭ-ma Piao, a scholar and member of the imperial family who lived between AD 240 and 305 (see for example Han,1955). Szŭ-ma Piao was attached to the imperial library and would thus presumably have free access to the official documents of the previous dynasty. The remainder of the history (annals and biographies) was written more than a century later by Fan-yeh (AD 398-445) and the whole was put together about AD 510. It is unfortunate that only very minor sections of the Hou-han-shu (including the astronomical records) have ever been translated into English.

Emperor Hsiao-ling, who was on the throne when the guest star occurred, reigned from AD 168 to 189. His capital was Lo-yang, which had been the seat of government since the beginning of the Later Han dynasty. As discussed in Chapter 2, we can confidently assume that the observations of the star which are incorporated in the extant record were made at Lo-yang ($34^{\circ}.7$ N) by the imperial astronomers. Hsiao-ling was the penultimate ruler of the dynasty, which at the time of his reign was already on the verge of collapse. The year before the star appeared, a major rebellion occurred in the northern provinces. This was one of the many calamities which brought about the downfall of the Han dynasty. The star thus appeared at a very critical time in Chinese history, and, like other celestial portents, must have been regarded as a harbinger of misfortune.

The imperial astronomers at this epoch were skilled observers as their records contained in Chapters 21, 22 and 28 of the Hou-han-shu testify. As an example, we might cite the following description of a meteor which occurred in the year AD 178: "1st year of the Kuang-ho reign period (of Hsiao-ling), 4th month, (day) kuei-ch'ou, a meteor (liu-hsing) trespassed against the 2nd star of Hsüan-yüan. It travelled north-east and entered the 'ladle' of Pei-tou".

The date of this event corresponds to AD 178 May 8. Hsüan-yüan lies in Leo and Lynx, but the actual star specified is uncertain. Pei-tou is the well known Plough, and the four stars forming the 'ladle' are α, β, γ, δ U. Maj. Several careful descriptions of the motions of comets which occurred during this same reign are given. The dates of these events are AD 178, 180, 182 and 188; in the latter year two separate comets are recorded. Translations of these observations are given by Ho Peng Yoke (1962) in his catalogue of comets and novae recorded in Far Eastern history.

One new star and five comets sighted in a period of 21 years is something of an achievement even by modern standards. However, planetary observations recorded in the Hou-han-shu are few and far between. There can be little doubt that only a minute proportion of the original observations of this type have survived. During Hsiao-ling's reign there are only five references to Venus, four to Mars and one to Jupiter; Saturn seems to have gone missing! (Mercury is, of course, very difficult to observe). Further, all but one of the planetary observations are of a very trivial nature, involving the passage of planets through various asterisms. Numerous essentially similar phenomena must have been noticed during this same period, but for one reason or another these have not found their way into the history. It might be remarked that this same feature, in varying degrees, is characteristic at all periods in Far Eastern history.

Despite the apparent dilatoriness of Hsiao-ling's astronomers in so far as planetary phenomena are concerned, each observation that *is* reported was regarded as of very great omen value, and is accompanied by a lengthy astrological commentary (*post factum*). Were these few observations selected because the astrological prognostications conformed to a basic pattern - foretelling the downfall of the dynasty? But this is a subject on its own ..... (for a useful discussion see Bielenstein, 1950). The possibility of falsification of astronomical records for astrological purposes cannot be ignored (see Chapter 2) and this might cast doubt on the authenticity of the account of the guest star, which so much concerns us here. With this in mind, we have felt it advisable to check the reliability of the planetary data by back calculation based on modern orbital elements. If the observations prove to be reliable, then we can obtain some idea of the accuracy which was achieved in astronomical observations.

The eight separate planetary observations which are recorded in the astronomical treatise of the *Hou-han-shu* during the reign of Hsiao-ling are as follows :

(a). "1st year of the Chien-ning reign period, 6th month, Venus was in the west. It entered *T'ai-wei* and trespassed against the star at the south end of the western 'wall'". The date corresponds to AD 168 July 23 - August 20.

(b). "1st year of the Hsi-p'ing reign period, 10th month, Mars entered within (*chung*) *Nan-tou*". Date : AD 172 November 4 - December 3.

(c). "2nd year (of the Hsi-p'ing reign period), 8th month, (day) *ping-yin*, Venus trespassed against (*fan*) the front star of *Hsin*". The year is equivalent to AD 173, but there was no *ping-yin* day in the 8th month.

(d). "5th year (of the Kuang-ho reign period), 4th month, Mars was within *T'ai-wei* and guarded (*shou*) *P'ing*". Date : AD 182 May 21 - June 18.

(e). "(5th year of the Kuang-ho reign period), 10th month, Jupiter, Mars and Venus, all three, met at *Hsü*. They were distant from one another 5 or 6 'inches' (*ts'un*) like a string of beads". Date : AD 182 November 14.

(f). "3rd year (of the Chung-p'ing reign period), 4th month, Mars retrograded and guarded the rear star of *Hsin*" . Date : AD 186 May 7 - June 4.

(g). and (h). "6th year (of the Chung-p'ing reign period), 8th month, (day) *ping-yin*, Venus trespassed against the front star of Hsin; (day) *wu-ch'en*, it trespassed against the middle star of *Hsin*". Dates : AD 189 September 20 (g) and September 22 (h).

In the reduction of the above terse, but nevertheless fascinating observations, we have used the planetary tables of Tuckerman (1964) to interpolate the positions of the planets. We have calculated the star positions from the data given by Newcomb (1910). Let us consider each observation briefly in turn :

(a). From the star maps of Ho Peng Yoke (1962), the source of most of our star identifications, the star at the south end of the western wall of *T'ai-wei* is clearly σ Leo. Calculation shows that on the evening of AD 168 July 23 - the 1st day of the 6th lunar month - Venus was only about 0.35 deg. from σ Leo. and crossed into *T'ai-wei*. The observations is thus highly accurate. We cannot say whether the day of the month was forgotten or lost at some later period.

(b). The constituent stars of Nan-tou are the bright stars μ, λ, φ, σ, τ, ξ Sag. During the 10th month Mars was much to the west of Nan-tou, but in the 1st month of the following year (i.e. 3 months later) the planet moved through the asterism, passing between μ and λ Sag. at the beginning of the month. The description 'within' Nan-tou would thus be accurate. Further discussion is left until after treatment of the remaining observations.

(c). The stars of Hsin are well established as the bright stars σ Sco. ("front star"), α Sco. ("middle star") and τ Sco. ("rear star"). As stated above, there was no ping-yin date in the 8th month so that we must leave the historical date in doubt. However, calculation shows that on the evening of AD 173 September 19 - the day hsin-ch'ou in the 8th month - Venus was less than 0.15 deg. from σ Sco. The month is thus correct, but for some reason or other the day is completely in error.

(d). During the first half of the 4th month, Mars was just inside the well defined southern boundary of T'ai-wei (determined by the stars σ Leo., β Vir. and η Vir.), and passed less than 5 deg. to the south of the asterism P'ing, whose constituents are ξ, ν, π and o Vir. The recorded description is thus accurate enough, although the planet was not stationary, as use of the term 'guarded' (shou) might imply.

(e). The three planets Jupiter, Mars and Venus were several degrees apart during the 10th month. However, continuing into the 11th month, there was indeed a very close triple conjunction. This is represented diagrammatically in Fig. 5.1, which shows the relative positions (longitude and latitude) of the planets on the evenings of AD 182 December 21 to 23. The configuration on December 22 (the 10th day of the 11th month) agrees splendidly with the graphic description of the text, "like a string of beads", and furthermore the planets were within the lunar mansion Hsü, as the record states.

(f). Mars was certainly retrograding during the 4th lunar month, but it did not approach close to τ Sco., the rear star of Hsin, until the following month. The planet's stationary point, reached in the middle of the 5th month, was less than 0.5 deg. to the west of τ Sco., but about 3 deg. to the north of it, so that for practically the whole of the 5th month Mars would appear to be suspended above the star, and could thus accurately be described as "guarding" it.

(g). and (h). On the evening of AD 189 September 20, Venus was less than 0.35 deg. from σ Sco., Two days later, the planet was only about 0.20 deg. from α Sco. The stated dates are thus correct and both observations are particularly accurate.

From the above analysis, it is clear that there is no evidence whatever of deliberate fabrication of astronomical observations during Hsiao-ling's reign. Numerous celestial events may have been omitted - deliberately or otherwise - but those which are recorded are of high reliability, and allow us to place considerable confidence in the account of the guest star of AD 185. The planetary records fall into two categories - those for which both the month and day are stated, and those for which only

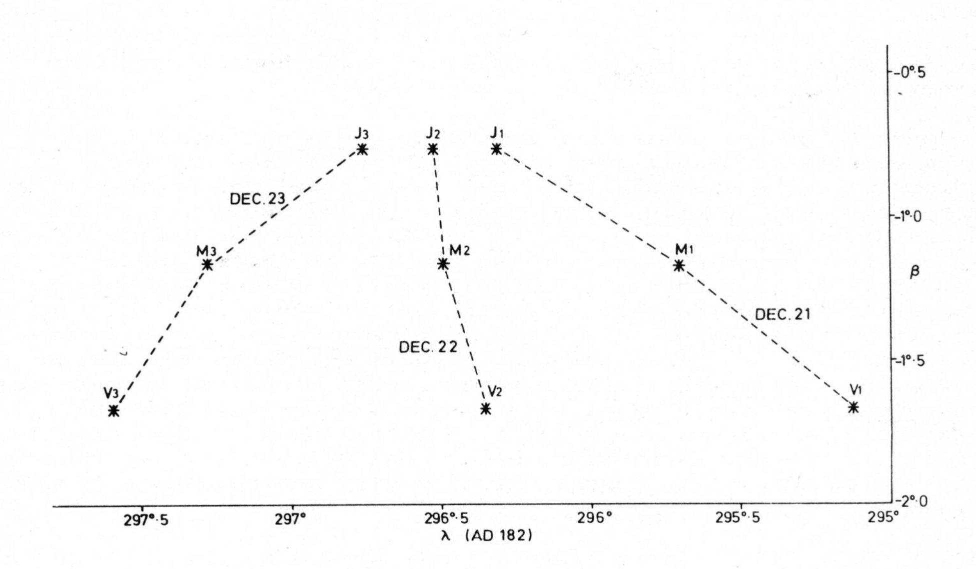

Fig. 5.1. Triple conjunction of Jupiter, Mars and Venus AD 182 December 21-23. Positions shown are for 1 hour after sunset.

the month is given. Of the former we have found two instances where the date is exact and a third where the month is correct but the day is completely erroneous. Where only the month is given, this is more often than not incorrect, but normally no more than a month in error. If errors of this amount can occur, it seems reasonable to assume that in the case of the observation listed as (b) above, the suggested error of 3 months is real.

Returning to the new star of AD 185, there is no record of it elsewhere in the history (Harvard - Yenching Index No. 41, 1966). However, as Fan-yeh, the compiler of the rest of the Hou-han-shu, wrote long after Szŭ-ma Piao, it is possible that the latter had access to material which was not available to Fan-yeh. The date of first appearance of the star, which is given to the day, is probably accurate; as noted earlier this corresponds to AD 185 December 7. However, only the month of disappearance is stated, and, as we have just seen, this should be treated with caution.

The duration of visibility has been variously interpreted. Whether it is approximately 8 or 20 months turns on the interpretation of the term hou-nien in the text. At the beginning of this chapter we have translated this as 'the year after next', implying the longer duration and a date of disappearance between AD 187 July 24 and August 21. However, Ho Peng Yoke (1962) renders the expression as 'the following year' and gives the date when the star ceased to be visible as AD 186 July 5 to August 2.

Hsi Tsê-tsung (1955) originally preferred this latter interpretation, but in his later paper in collaboration with Po Shu-jen (1965) he favours the longer duration. The period of visibility of the star is so important that it is worth considering this seemingly minor technicality in some depth.

In modern Chinese usage the expression *hou-nien* always means 'the year after next', just as *hou-jih* means 'the day after tomorrow'. The equivalents of our 'next year' and 'tomorrow' are respectively *ming-nien* and *ming-jih*. On the early use of *hou-nien* the dictionary of Morohashi (1955) is not too definite, and in order to ascertain whether the meaning in later Han times (or at least when the astronomical treatise of the *Hou-han-shu* was written) was the same as at present we did a count of the relative frequency of occurrence of *ming-nien* and *hou-nien* in the remainder of the astronomical treatise. There are twelve examples of *ming-nien*, but only this single isolated example of *hou-nien*. Frequently *hou-x-nien* (where x is a small number) is also used, meaning x years later, but the usage of *hou-nien* itself is unique, suggesting that it was deliberate. We thus feel that we can adopt the longer duration with a high degree of confidence.

A period of visibility as long as some 20 months makes the probability that the star was a supernova extremely high; a comet can be entirely ruled out. It should be pointed out that a duration of more than a year implies that the text remains silent on the disappearance and recovery of the star around the time of conjunction with the Sun (heliacal setting and rising). However, this is not a major difficulty. The supernova of AD 1054 is far better documented than the star under consideration (see Chapter 8 below), but although both the dates of first sighting and final fading are given, corresponding to a period of visibility of nearly two years, there is not a single reference to heliacal setting or rising.

The asterism *Nan-mên*, in which the guest star appeared, was probably the southernmost star group visible from central China. In the astronomical treatise of the *Chin-shu* we read "The two stars of *Nan-mên* ("Southern Gate"), situated south of *K'u-lou*, form the outer gate of the heavens and govern garrison troops" (trans. Ho Peng Yoke, 1966). The importance of this description lies in its assertion that *Nan-mên* consisted only of two stars. Ho Peng Yoke in his commentary on the above quotation was of the opinion that the two components were α and ε Cen., but in his star maps at the end of his book he links β and ε Cen. Hill (1967) in his pioneering discussion of the possible radio remnants of the new star of AD 185 preferred to leave the question open, merely stating that "*Nan-mên* was apparently associated with the stars α, β and ε Cen". However, the triangle formed by these three stars covers a fairly large area of sky (some 10 square degrees). In order to clarify the issue, Stephenson (1975) undertook a careful examination of copies of oriental star maps. The originals of these vary in date from about AD 900 to 1750. This investigation has since been extended to cover the very early (around 2,000 years old) star maps in the

Hsing-ching ("Star Manual"). All of these charts are agreed that Nan-mên consisted only of two stars, and although most are crudely drawn, estimates of the right ascension and declination of the asterism (especially from the later maps) are sufficient to prove that we must indeed choose from α (mag. -0.3), β (mag. +0.6) and ε Cen. (mag. +2.3). There are no other bright stars in the vicinity (all others are fainter than mag. +4), and, as the asterism was only a few degrees above the horizon in Central China, no other nearby stars would be expected to be readily visible.

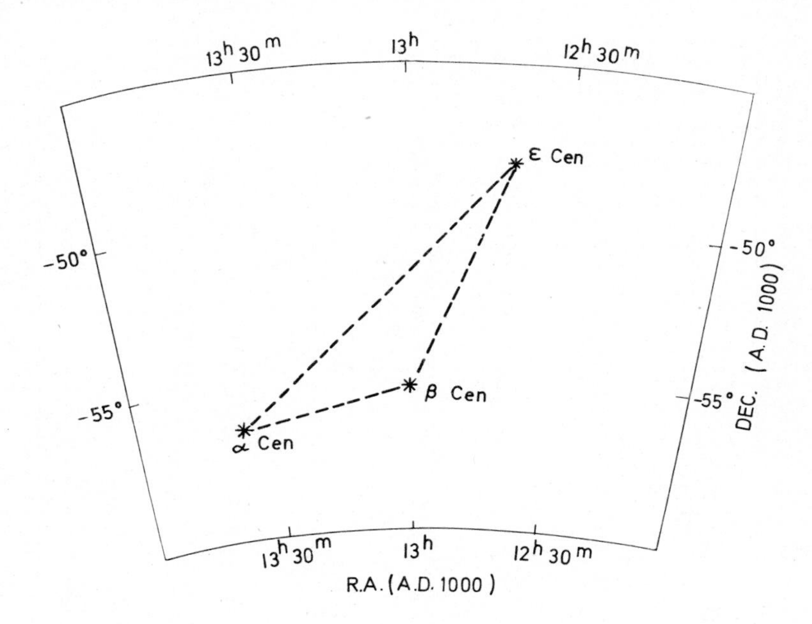

Fig. 5.2. Positions of α, β and ε Cen. at epoch AD1000.

The celestial co-ordinates of α, β and ε Cen. are shown diagrammatically in Fig. 5.2 for AD 1,000 - a typical mean epoch for comparison with the representation of the asterism on the various oriental star maps. The figure should be compared with Fig. 5.3. This is an illustration of a star map in the Ku-chin-t'u-shu-chi-ch'êng, a Chinese encyclopedia dated from AD 1725, but containing star charts (pre - Jesuit) from more than a century earlier. The region of sky covered is roughly from about 10^h to 14^h right ascension and from -60° to $+50^\circ$ declination. This chart is almost unique among oriental representations in that it makes some attempt to distinguish between stars of different brightness. Three classes are recognised. Undoubtedly the three bright stars are α Vir. (near the centre of the chart) and α and β Cen. near the lower edge. It is evident that the map is not drawn to scale (a common failing), but nevertheless there can be no question that Nan-mên was understood to consist only of α

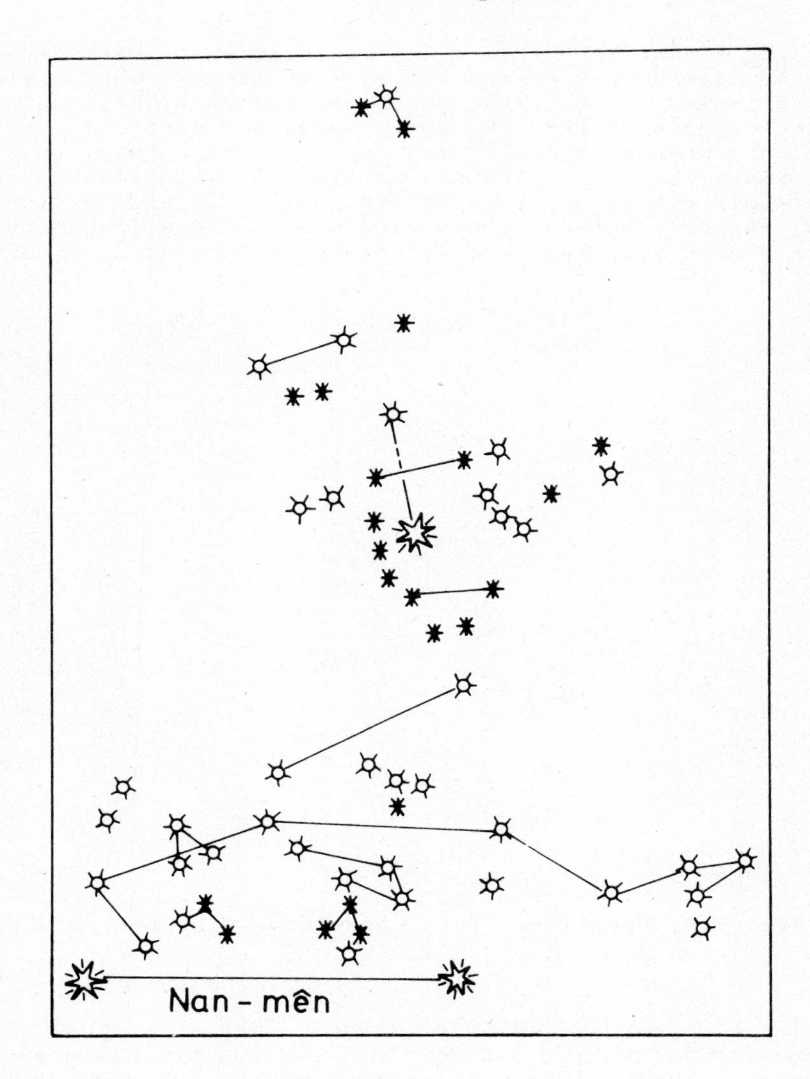

Fig. 5.3. A copy of an early 17th century star chart showing a small section of the sky. We have omitted the Chinese characters identifying the various asterisms and inserted the romanisation for Nan-mên, where SN 185 appeared. The three bright stars are α Vir. (near the centre of the map) and α and β Cen., forming Nan-mên.

and β Cen. (there is an isolated star further north which could possibly represent ε Cen.).

The other oriental star maps which we have consulted do not differentiate between stars of different brightness, cf. Plates 3 and 4. In order to determine the components of Nan-mên, we found the most satisfactory technique was to measure on each chart the angle between the line joining the two stars and an estimated circle of declination passing between them. This angle was in all cases between 0° and $+35^{o}$ (measured towards the east), with a mean of $+15^{o}$. In Fig. 5.3 it is close to 0°, but in Plate 4 it is about $+30^{o}$. This is in acceptable agreement with the actual angle for α and β Cen ($+10^{o}$), but incompatible with the very large angles for α and ε Cen or β and ε Cen. Accordingly we have no hesitation in identifying α and β Cen as the two stars of Nan-mên. This considerably reduces the area of search.

Let us now consider the position of the new star in relation to Nan-mên. It was described as "within" (chung) the asterism. The usual expression to denote the location of a planet or new star is the rather vague term yü, which may be translated "at" or "in the vicinity of". However, chung, which is comparatively rare, is more specific. Depending on the context, it can be rendered as "middle" or "within". In its former capacity it is used in the name for China itself - Chung-kuo ("The Middle Kingdom"). There are various examples of the latter meaning in the astronomical treatise of the Hou-han-shu. Thus we find several instances in Chapters 21 and 22 where a planet is said to have entered within an asterism. An occultation of Venus in AD 72 is described as "Venus entered within the Moon" (Chapter 21). Sunspots in AD 188 and 189 were identified as follows: (AD 188) "a black vapour as large as a melon was within the Sun"; (AD 189) "the Sun was orange in colour and within it there was a black vapour like a flying magpie. After several months it melted away".

In order to check the precise significance of chung where an asterism is concerned, we examined the various planetary observations recorded in the Hou-han-shu. In most cases the planets were said to enter within star formations containing a number of rather faint stars, making their identification questionable. However, two observations, both involving the planet Venus seemed very reliable since the star group concerned (the same in each case) was bright and the location within it was given. In both AD 117 and 125 Venus is reported as entering within the "mouth" (k'ou) of Nan-tou. This well-defined lunar mansion in Sagittarius, whose name may be translated as "Southern Dipper" is very similar in form to the better known "Northern Dipper" or "Plough" in Ursa Major. The constituent stars are readily identified as φ, λ, μ, σ, τ and η Sgr. for there are no other bright stars in the vicinity.

The dates of these two Venus observations, which are recorded in chapter 21 of the Hou-han-shu, correspond to AD 117 October 23 and AD 125 October 24. It is perhaps noteworthy that these dates are almost exactly 8 years apart (i.e. 5 synodic periods of Venus) but it is doubtful if the Chinese astronomers were aware of the Venus cycle of this length. Fig. 5.4 shows the positions of the stars of Nan-tou at the mean epoch AD 121, corrected for precession and proper motion. The two locations of Venus are calculated for the evenings of AD 117 October 23 (V1) and AD 125 October 24 (V2), and the approximate motion of the planet in one day is shown for comparison. There can be no doubt that the "mouth" of Nan-tou is formed by φ, σ, τ and ε Sgr. From Fig. 5.4, as well as the evidence that both recorded dates are exactly correct, it is clear that the observations are accurate to within a small fraction of a degree. On each occasion Venus was just entering the "mouth" of the asterism. Chung was indeed a precise term.

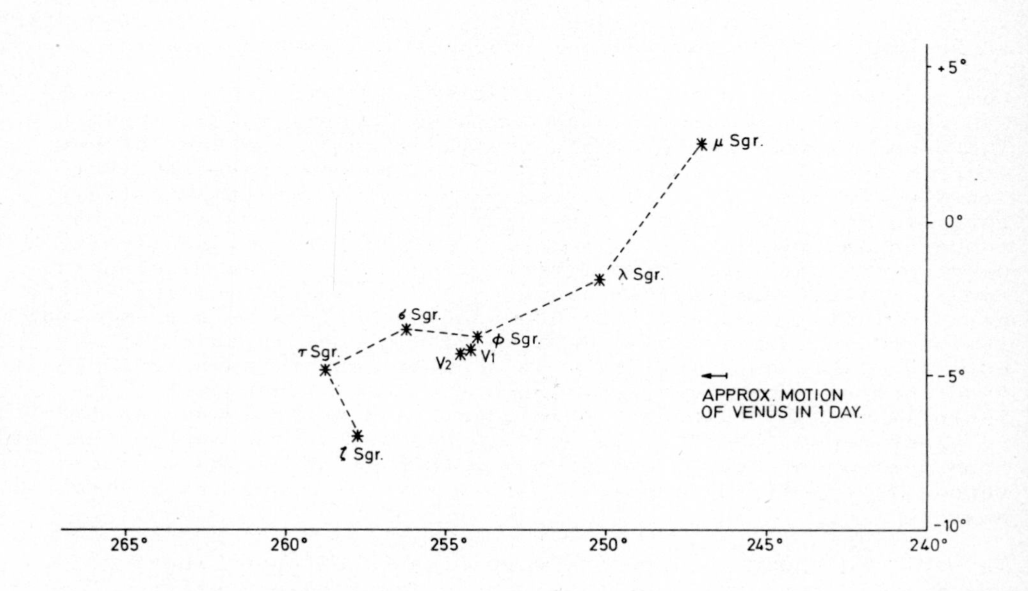

Fig. 5.4. Observations of Venus entering within the 'mouth' of the asterism Nan-tou, AD 117 October 23 (V1) and AD 125 October 24 (V2). Positions shown are for 1 hour after sunset.

In the context of the guest star record, we can now place the star roughly between α and β Cen., and in looking for its remnant consider only those SNRs which answer this location.

The final observational question concerns the apparent brightness of the supernova. It was never far above the horizon at Lo-yang (the meridian altitudes of α and β Cen. were respectively only 3°

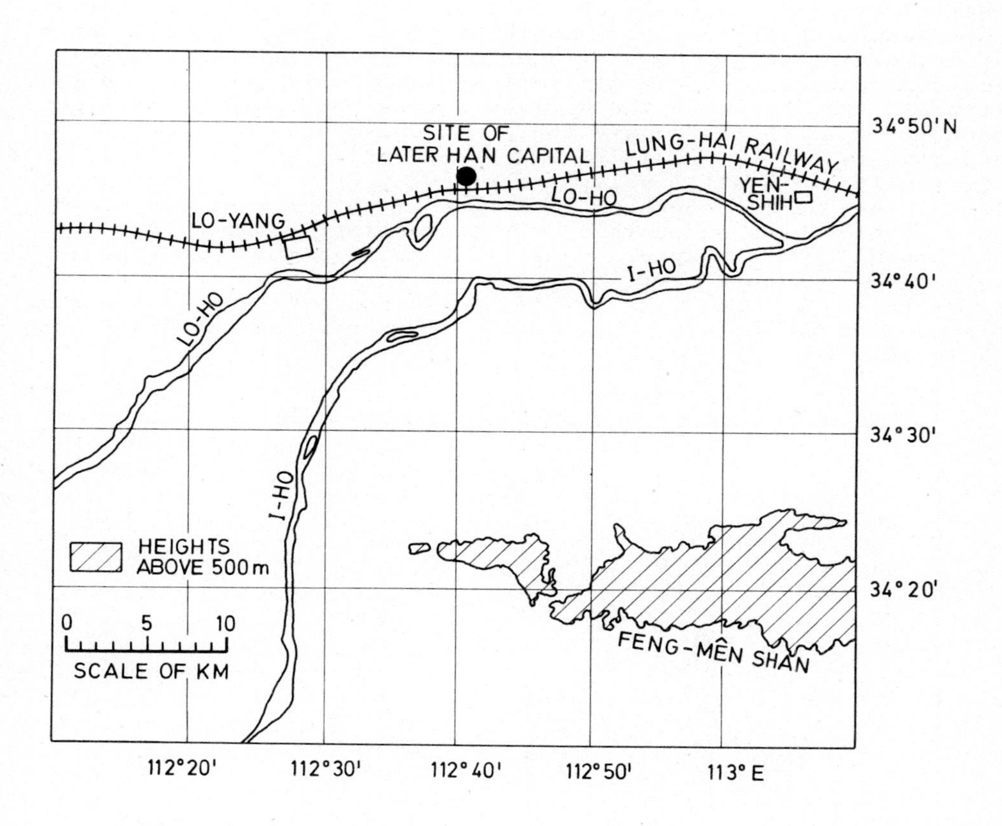

Fig. 5.5. A simplified map of the Lo-yang area.

and 5°) and would only be above the horizon for a very few hours each day. We have consulted a detailed (1 : 250,000) relief map of the Lo-yang area with contours at 100 m intervals. This is shown in simplified form in Fig. 5.5, with the site of the ancient capital indicated; to avoid unnecessary confusion only the 500 m contour is included. An horizon profile is shown as Fig. 5.6, and this indicates that the southern horizon from Lo-yang was relatively, unobscured (more later). At low elevation atmospheric absorption would be considerable, amounting to some 3 mag. Probably the optical dispersion and apparent physical size of the object (as described in the text) can be accounted for in this way. Because of the poor visibility of the star we must assume an error in the month of final disappearance. Even at the beginning of the 6th lunar month in AD 187 (late July) the star was setting almost simultaneously with the Sun, so that it would only be above the horizon during the hours of daylight. However, a month previously there would probably be a brief opportunity to observe it in a dark sky after sunset. (N.B. This difficulty would still arise even if we preferred a duration of less than a year). Several instances of an error of one month have already been encountered - the planetary observations discussed in this chapter - and it seems legitimate to suppose this to be the case here, although, of course, we can never prove this. We thus make the tentative suggestion that the star disappeared in the 5th lunar month, corresponding to AD 187 June 24 to July 23.

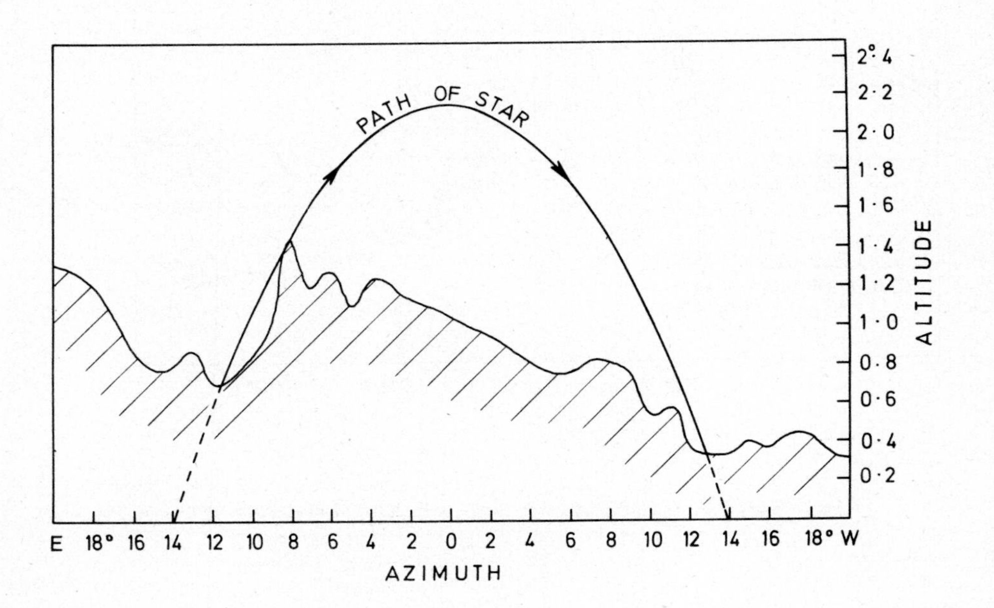

Fig. 5.6. Profile of the southern horizon from Lo-yang showing the apparent diurnal path of a star having the same location as G315.4-2.3.

In order to obtain some preliminary estimate of the apparent magnitude of the star at maximum we can use the method outlined in Chapter 1. This makes use of the duration and assumes typical Type I supernova parameters. Taking the final magnitude as +2 (allowing about 3 mag. for atmospheric extinction) gives a rough estimate for the magnitude at maximum of about -8. The extraordinary brilliance is highlighted by the fact that, if the stated date of appearance of the guest star is correct, its visibility must have been seriously impaired by the dawn-glow, since the star was rising just ahead of the Sun (more later). A distance in the region of 1 to 2 kpc thus seems to be indicated. We therefore seek a nearby remnant of age 1800 years lying roughly between α and β Cen. Clark and Caswell's (1976) catalogue of SNRs lists four remnants in this area of Centaurus : namely G311.5-0.3, G315.4-2.3, G316.3-0.0, and G315.4-0.3. This region has now been particularly well surveyed, and there are no other sources which could be considered as possible SNRs. Fig. 5.7 shows the position of the four candidate SNRs with α and β Cen. for epoch AD 185. (Proper motions of the SNRs have been ignored; however this is unlikely to be the cause of significant error.) As noted in Table 4.2, the only SNR in this region that appears to be near enough and of small-enough diameter (i.e. young enough) to be the remnant of AD 185 is G315.4-2.3. Nevertheless, to confirm this initial assessment we will consider here all four SNRs of Fig. 5.7.

G311.5-0.3 was first identified as an SNR by Shaver and Goss (1970) from their partial southern Galactic Survey, and it was included as Number 35 in the early SNR catalogue of Downes (1971). The source is of small angular diameter and is confused with an adjacent H II region (G311.5-0.5). Presently available radio

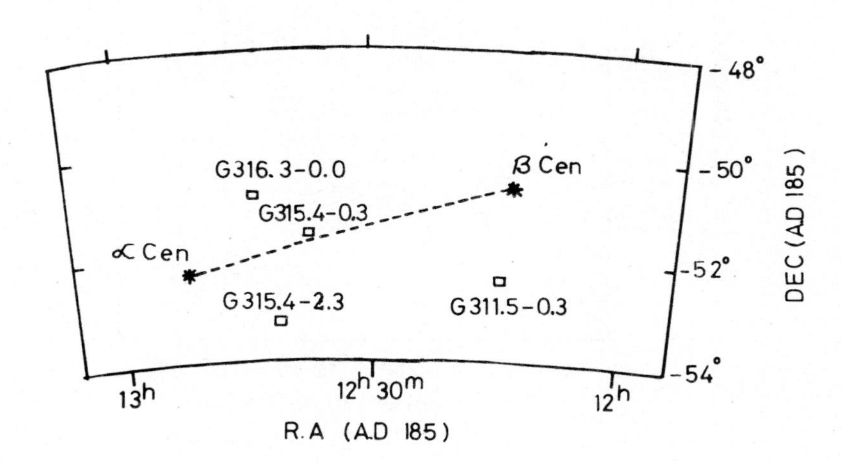

Fig. 5.7. The position of the four candidate SNRs for the supernova of AD 185, with α and β Cen.

maps have insufficient resolution to detect peripheral brightening. The H I absorption technique described in Chapter 4 for distance determination gives a minimum distance estimate of 6.6 kpc (Caswell et al, 1976); in fact the small angular diameter and a low radio-frequency flux density measured for the source suggest that it is probably on the remote side of the Galaxy. The large inferred distance for G311.5-0.3 makes it a highly unlikely candidate for the remnant of the supernova of AD 185.

G316.3-0.0 was catalogued as MSH 14-57 in the pioneering 85 MHz southern radio survey of Mills, Slee, and Hill (1957), and was included in the Milne (1970) catalogue of SNRs. High-resolution maps show the double-lobe configuration characteristic of an SNR viewed across the direction of the interstellar magnetic field (see Chapter 4.) A previously unpublished radio map of the source made by the authors with the Molonglo radiotelescope at 408 MHz is shown as Fig. 5.8. An H I absorption distance estimate (Caswell, 1967) places this SNR at a distance of $\geqslant 7.2$ kpc with diameter $\geqslant 36$ pc. This extreme distance estimate must preclude G316.3-0.0 from consideration as the remnant of the supernova of AD 185.

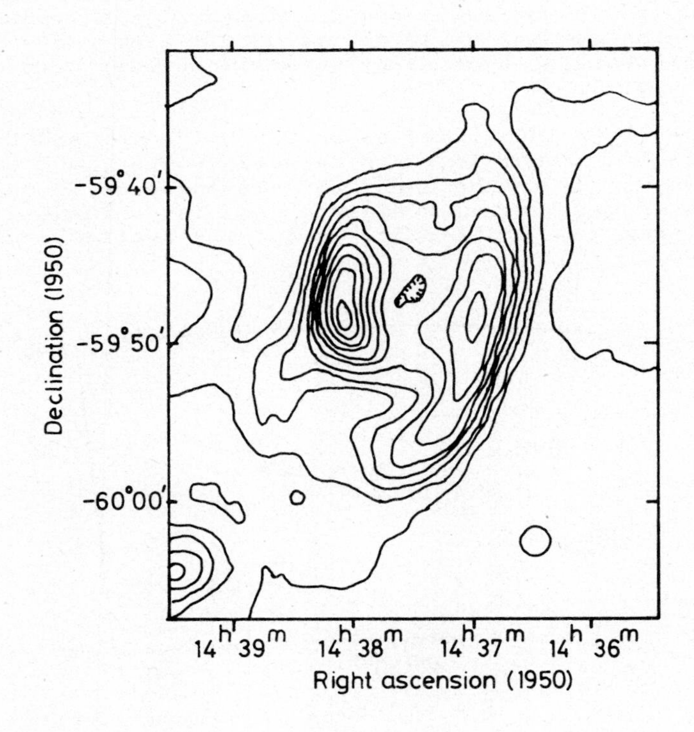

Fig. 5.8. A 408 MHz map of the SNR G316.3-0.0.

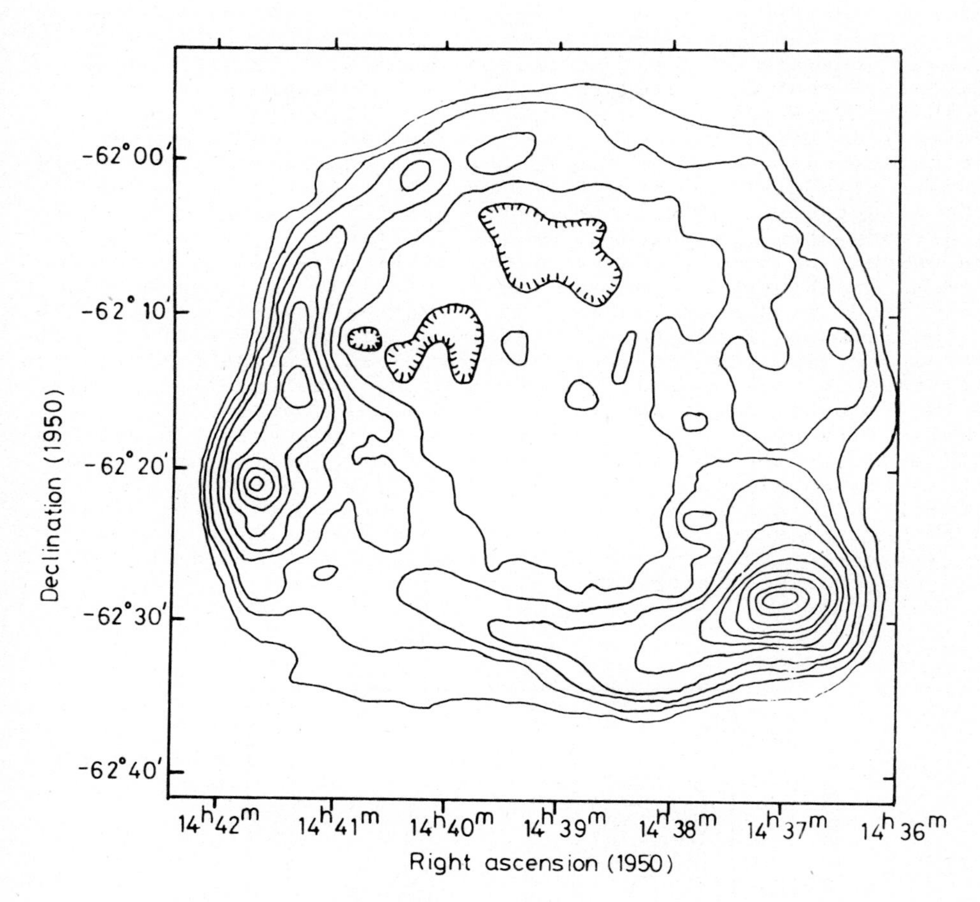

Fig. 5.9. A 408 MHz map of the SNR G315.4-2.3, the likely remnant of the supernova of AD 185.

The radio source G315.4-2.3 was also present in the 85 MHz catalogue, as MSH 14-63. Hill (1964) was the first to present evidence of its possible supernova character, and he later published the results of observations at 1410 and 2650 MHz clearly indicating a non-thermal spectrum and peripheral brightening (Hill, 1967). In Fig. 5.9 we show the highest resolution radio map presently available compiled from observations made by the authors with the Molonglo radio telescope at 408 MHz. The source is 40 arc min. in angular diameter, and shows a well-defined broken shell structure. Comparable resolution observations at 5000 MHz made with the Parkes radiotelescope (Caswell, Clark and Crawford, 1975) show a similar structure, and the 408 MHz and 5000 MHz flux density estimates yield a non-thermal spectral index of -0.62. Milne (1972) detected significant polarization from the regions of peak radio brightness, commensurate with the non-thermal nature of the source. Optically the remnant consists of a set of bright filaments, designated as RCW 86. A weak elongated filament lies along the northern edge of the radio shell, and a set of bright filaments (shown as Plate 7) coincides with the region of enhanced radio emission in the south-west corner of the remnant. A concentration of early B stars at a distance of 2.5 kpc was the basis of Westerlund's (1969) suggestion of this being the possible distance to RCW 86. As already noted, a somewhat smaller distance is preferred for the remnant of the supernova of AD 185 because of its probable extreme brightness at maximum. This possible distance discrepancy is of little consequence in the consideration of G315.4-2.3 as the remnant of the AD 185 supernova, since both distance estimates are highly subjective. Nevertheless, on the basis of the brightness of the optical filaments alone, the relative proximity of the remnant to the Sun is beyond dispute, making G315.4-2.3 a highly likely candidate for the AD 185 supernova. In addition, G315.4-2.3 is the southernmost of the four SNRs in the region, with meridian altitude at Lo-yang about 2.2° (allowing for refraction). A star this close to the horizon would be likely to suffer the severe dispersion ("it showed the five colours") and distortion ("it was as large as half a mat") described in the text for the guest star.

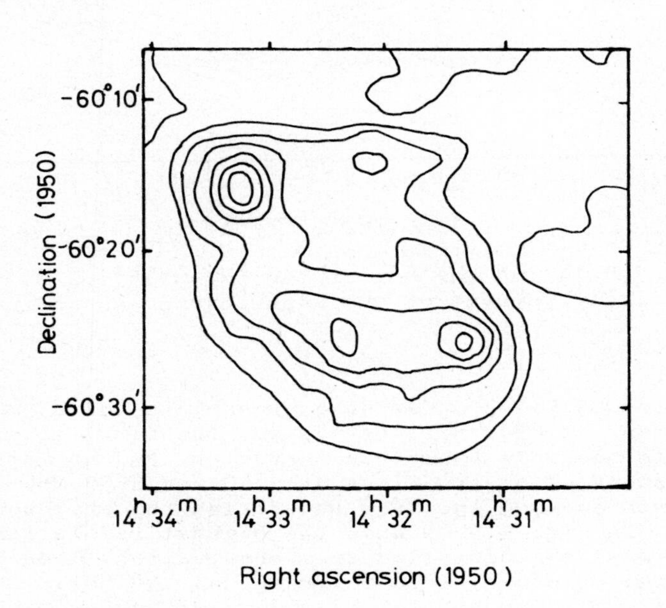

Fig. 5.10. A 408 MHz map of the SNR G315.4-0.3.

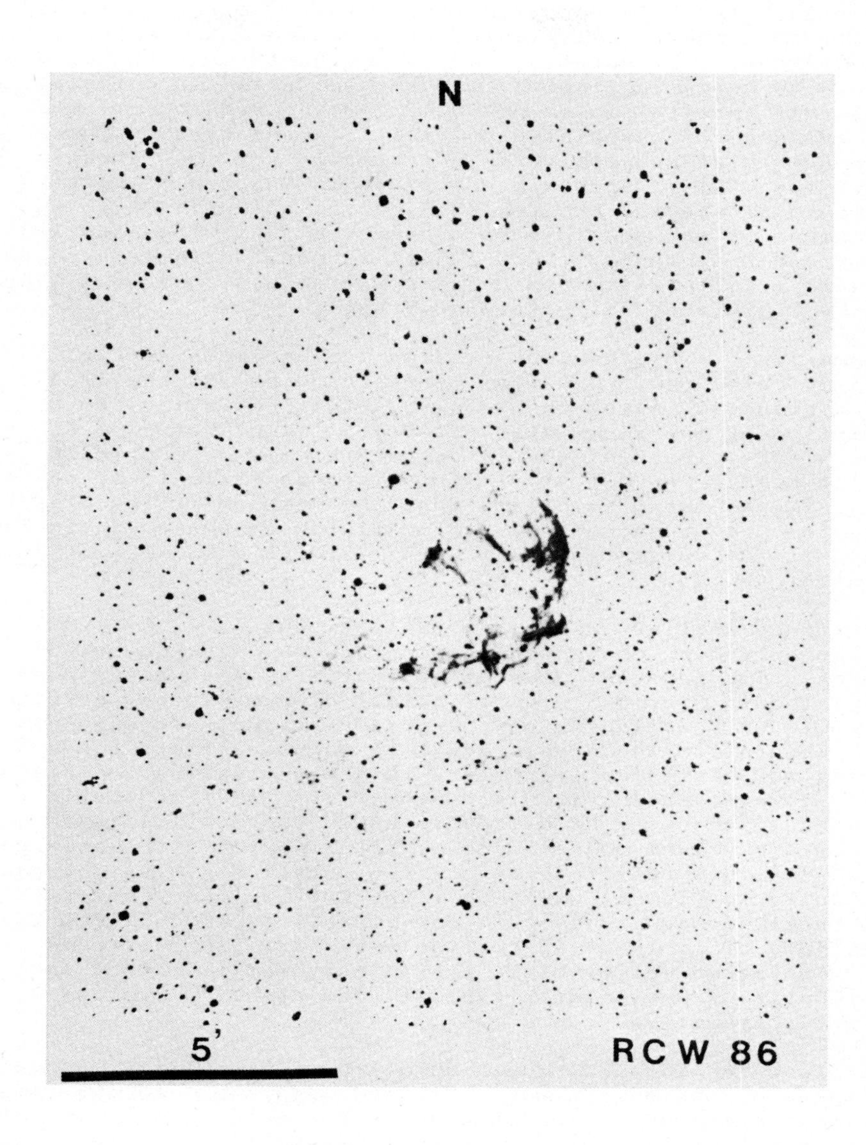

Plate 7. The optical remnant RCW 86.

We come now to the final SNR to be considered, G315.4-0.3. This is one of 28 southern radio sources recently recognised to be SNRs by Clark, Caswell and Green (1975). The Molonglo 408 MHz map for G315.4-0.3 is shown in Fig. 5.10. The source is ≃16 arc min angular diameter, displaying an enhanced emission in the south-west quadrant. The point source on the north-eastern boundary is believed to be extragalactic, and was extracted in calculating a non-thermal spectral index of -0.47 for the remainder. There is no evidence of associated optical filamentary structure. No independent distance estimate is available for the remnant. However the "Σ-D" approach described in Chapter 4 suggests a distance of about 8 kpc for G315.4-0.3, but with large uncertainty since it may possibly be a nearby object with anomalously low surface brightness. (RCW 103 is recognised to be such an SNR, with a well-determined H I absorption distance of 3.3 kpc, but a Σ - D distance estimate of 8.7 kpc).

To summarize : of the four candidate SNRs shown in Fig. 5.7, G311.5-0.3 and G316.3-0.0 must be rejected on the basis of their large distances. Although G315.4-0.3 fits perfectly the Chinese description of the guest star of AD 185 lying 'between' (*chung*) α and β Cen., its distance inferred from a Σ - D approach is much too high ; an H I absorption distance estimate is needed to investigate the possibility that the remnant may be closer. In the absence of such data, the available evidence continues to support the suggestion that G315.4-2.3 (RCW 86) is the *probable* remnant of AD 185.

As mentioned earlier, a star in the position of G315.4-2.3 would be only 2.2° above the horizon at Lo-yang when on the meridian. The apparent diurnal path of a star at this location is depicted in Fig. 5.5. Under normal circumstances, a bright star would be readily visible at this altitude, although atmospheric absorption would be heavy - as much as 4 magnitudes (at very low altitudes light loss in the atmosphere increases rapidly towards the horizon). However if G315.4-2.3 is in fact the remnant, the star at discovery would be to all intents and purposes a daylight object. Before sunrise the star would be scarcely visible (altitude at sunrise only 1.7°), and the best opportunity to view it would be after the Sun had risen. Assuming a required magnitude of -4 for visibility in daylight leads to a magnitude of at least -8 for the new star. This extreme brightness is in agreement with the preliminary estimate made earlier on the basis of period of visibility and assuming typical Type I supernova parameters.

Our preferences for a 20 month duration of visibility, an extreme brightness at maximum (≃ mag. -8), and G315.4-2.3 as the remnant, do in fact allow an interpretation of the supernova of AD 185 as either Type I or Type II. If Type I (assuming visual absolute magnitude at maximum -19), allowing an interstellar absorption of 1 mag. / kpc gives a source distance of ≃ 1 kpc, a remnant diameter of ≃12 kpc, and an average expansion velocity of ≃6,500 km/sec. The remnant would be at greater distance if the estimate of maximum apparent magnitude is high, although a distance as great as 2 kpc would be possible only if one were to assume an apparent magnitude of -5; as noted above such a value appears to be incompatible with the supernova's inferred day-time sighting at low meridian altitude. If Type II (assuming visual absolute magnitude at maximum -17.5), the corresponding parameters are distance ≃600 pc, diameter ≃7 pc, and average expansion velocity ≃4,000 km/sec. In principle either set of parameters would be acceptable, and there is clearly nothing in the historical record which would enable us to distinguish between Type I and Type II. Nevertheless, and this judgement must be conjectural in the absence of convincing experimental evidence of the behaviour of Type II light curves beyond about 150 days, the long period of visibility tends to favour a type I interpretation.

Since the guest star of AD 185 is the oldest supernova recorded historically, we have felt justified in subjecting the single text referring to it to close scrutiny. In particular, it was believed necessary to test the reliability of the positional description and the recorded duration of visibility, since an unambiguous identification of the remnant of the supernova depended on this information. Despite minor errors in the dates recorded for certain astronomical events, there can be no doubting the essential reliability of the Chinese observations at this time. Their single record of the supernova of AD 185, and the identification of its probable remnant based on the historical description, will prove invaluable in investigating the evolution of supernova remnants up to an age of 2000 years.

Chapter 6

THE CHIN DYNASTY GUEST STARS

Continuing in chronological order, the next two new stars which rate as possible supernovae were observed in AD 386 and 393. As both stars were recorded only by the Chinese, and occurred in the same reign period, it seems appropriate to consider them together, especially as each description is very brief.

Both stars appeared during the latter years of the Chin Dynasty in China, but the earliest records of them which we possess are contained in a section of the astronomical treatise (Chapter 25) of the Sung-shu. The entire history is the work of one man, Shen Yüeh, who lived between AD 441 and 513. The astronomical treatise contains many observations dating from the Chin period.

After Shen Yüeh completed the Sung-shu, more than a century was to elapse before the compilation of the official history of the Chin Dynasty - the Chin-shu. From Ho Peng Yoke (1966), this latter work was compiled by a team of scholars under the edict and supervision of Emperor T'ai-tsung, the second ruler of the T'ang Dynasty. At that time there existed no less than 18 different versions of the history, but all of these were found to be inadequate to serve as an official history. The astronomical chapters of the Chin-shu were in all probability written by Li Shun-fêng (AD 602 - 670), an eminent astronomer and mathematician, but because he wrote so long after Shen Yüeh, it is likely that the latter had access to more direct sources than Li Shun-fêng.

The astronomical treatise of the Sung-shu follows the pattern of that in the Hou-han-shu : The various types of observation are not classified according to type. However, in the corresponding section of the Chin-shu, astronomical records are classified under as many as nine different headings. The new stars of AD 386 and 393 are to be found in the section entitled "Ominous stars and guest stars", which forms part of chapter 13 of the history.

Ho Peng Yoke's (1966) excellent translation into English of the astronomical chapters of the Chin-shu has made this work generally accessible in the West. No such translation of the Sung-shu exists, although in his commentary on the Chin-shu records Ho Peng Yoke frequently quotes from it.

The reports of the two new stars in both the Sung-shu and the Chin-shu are very much alike, both in the description of the stars and in the accompanying astrological commentary. We have felt it desirable to give translations of the accounts of the stars from both sources, but to avoid unnecessary repetition have quoted the astrological interpretations from the later

Chin-shu only.

AD 386 :
Sung-shu (Chapter 25). "11th year of the T'ai-yüan reign-period, 3rd month, there was a guest star at Nan-tou until the 6th month when it was extinguished (mieh)".
Chin-shu (Chapter 13). "11th year of the T'ai-yüan reign-period, 3rd month, there was a guest star at Nan-tou until the 6th month when it vanished (mo)".

The months of appearance and disappearance correspond to AD 386 April 15 - May 14 and July 13 - August 10.
Astrological Prognostication. "The interpretation was military activities and an amnesty. After this (the people of) Szŭ, Yung, Yen, and Chi were frequently called up for military service. Later in the 12th year, 1st month, there was a general amnesty, and in the 8th month another general amnesty".

AD 383 :
Sung-shu (Chapter 25). "18th year of the T'ai-yüan reign-period, 2nd month, there was a guest star within (chung) Wei until the 9th month when it was extinguished (mieh)".
Chin-shu (Chapter 13). "18th year (of the T'ai-yüan reign-period), 2nd month, there was a guest star within (chung) Wei until the 9th month when it vanished (mo)".

The months of appearance and disappearance correspond to AD 393 February 27 - March 28 and October 22 - November 19. (It should be noted that there was an intercalary 7th month in the Chinese calendar during this interval).
Astrological Prognostication. "The interpretation was military activities and deaths at Yen. In the 20th year Mu-jung Ch'ui and Hsi-pao attacked Wei, but were defeated and more than 10,000 men were killed. In the 21st year (Mu-jung) Ch'ui died, and eventually his state was ruined".

The great similarity between the records of AD 386 and AD 393 is self-evident, suggesting that the original reports were reduced to a stereo-type pattern, presumably by Shen Yüeh. The Chin-shu copies the style of the Sung-shu, changing only the final character (mieh to mo), but without altering the meaning in any way.

The imperial annals of the Chin-shu (Chapter 9) are silent on the new stars of AD 386 and 393 but in any case they contain few astronomical observations - mainly solar eclipses. We do not expect any additional records from the Sung-shu since the annals of this history only deal with the Liu-sung Dynasty itself. We have searched several of the early Chin histories for references to the guest stars, but without success.

It is perhaps something of a coincidence that like their counterpart in AD 185, the stars of AD 386 and 393 heralded the downfall of a dynasty. The Chin Dynasty was far weaker than the Han, and as early as AD 317 the western half of the empire was lost to barbarians. The capital was then removed to Chien-k'ang (modern

Nan-ching), where it remained until the fall of the dynasty in AD 420. Hsiao-wu, who was emperor at the time when the guest stars appeared, reigned from AD 373 to 397. He was a weak, incompetent ruler who did nothing to stem the tide of rebellion. After his death at the age of 34 (he was murdered while hopelessly drunk by his favourite concubine) the dynasty dragged on for a further 23 years before its final end. The astrological prognostications accompanying the various astronomical observations made during Hsiao-wu's reign reflect the troubled state of these times and this is particularly true of the guest star records. Two planetary observations at this time were taken as portents of the emperor's death - an occultation of Mars by the moon (AD 393 February 20), and a conjunction of Jupiter with two small asterisms (AD 396 July 22 - August 19).

It is unfortunate that the records of both new stars tell us so little of astronomical interest. No reference to brightness or colour is made and only the months of discovery and last sighting are given. Bearing in mind the political turmoil of these times, the complete fabrication of astronomical records and astrological prognostications under political pressures might seem to have been possible. Therefore, it seems desirable once again to test the reliability of astronomical records in general around this period, and in particular to check the accuracy of dates recorded only to the nearest month.

During the reign of Emperor Hsiao-wu, six observations relating to the motions of planets are reported for which the date is given only to the nearest month. All observations were made during the second part of his reign - the T'ai-yüan reign-period - in which the guest stars appeared. In the following translations of the observations the year is given in terms of the T'ai-yüan reign period, the first year of which corresponds to AD 376-377. The text used is the _Sung-shu_, but any significant discordances between this and the _Chin-shu_ will be commented upon.

(a). "(1st year) 9th month, Mars trespassed against _K'u-hsing_ and _Ch'i-hsing_ and then entered _Yü-lin_". Date AD 376 September 30 - October 28.

(b). "3rd year, 6th month, Mars guarded _Yü-lin_". Date AD 378 July 11 to August 9. The _Chin-shu_ gives the date as "2nd year, 2nd month", which corresponds to AD 377 February 24 to March 25.

(c). "(13th year, 12th month) Mars was at _Chüeh_ and _K'ang_". Date AD 389 January 13 to February 11.

(d). "14th year, 12th month, Mars entered _Yü-lin_". Date AD 390 January 3 to 31.

(e). "19th year, 10th month, Venus, Saturn, Mars and Mercury met at _Ti_". Date AD 394 November 10 to December 8. This observation is reported in the _Chin-shu_ but not in the _Sung-shu_.

(f). "20th year, 6th month, Mars entered _T'ien-chün_". Date AD 395 July 4 to August 1.

It will be noticed that each of the observations mentions the planet Mars, but there seems to be no special reason for this. There are plenty of sightings of Jupiter and Saturn during this same period, but here the precise date is given (this also applies to a number of further observations of Mars).

A brief discussion of each record is as follows :

(a). K'u-hsing, Ch'i-hsing and Yü-lin are all rather faint asterisms lying mainly in Aquarius. During the 9th month Mars was moving eastwards so that the first asterism reached would be K'u-hsing. This consists of two stars, of which the westernmost is μ Cap. As it happens, the planet did not make its closest approach to μ Cap. until the 3rd day of the 10th month. In this latter month Mars passed close to Ch'i-hsing and skirted the northern edge of Yü-lin. We have here an error of a month in the recorded date, but otherwise the motion of the planet was accurately described.

(b). As already pointed out, there is considerable discord between the dates recorded in the Sung-shu and the Chin-shu for this event. In reality, neither date is correct : during the two years in question, Mars was only in the vicinity of Yü-lin during the 3rd and 4th months of the 3rd year. There is an unaccountable error here.

(c). Chüeh, which consists only of the two bright stars α and γ Vir., is a particularly well-defined asterism, and K'ang, also in Virgo and lying some 10 deg. to the east of Chüeh, is also fairly bright. About the 20th day of the 11th month, Mars moving eastward, passed between the stars of Chüeh, but during the 12th month the planet was almost stationary between the two asterisms. The recorded month is thus probably correct.

(d). Yü-lin is an extensive asterism consisting largely of faint stars. It lies a little to the south of the ecliptic in Aquarius. During the latter half of the 11th month and the first day or two of the 12th month, Mars, moving rapidly eastward, passed along the northern edge of Yü-lin. In view of the difficulty in identifying several of the constituents of this asterism, it is not possible to say just when the planet actually entered the asterism (if at all). The recorded date thus may be correct or a month in error.

(e). This is perhaps the most interesting observation of the six. Four planets in close proximity to one another is a rather rare occurrence. Calculation shows that from the 15th to the 18th day of the 10th month (November 24 to 27) the planets Mercury, Venus, Mars and Saturn covered a range of no more than 14 deg. What is particularly important is that on the first two of these dates all four planets were within the lunar mansion Ti, in accord with the Chinese description. Here the recorded month is unquestionably correct.

(f). T'ien-chün, in Cetus, is a fairly well defined asterism lying a little to the south of the ecliptic. Commencing about the 20th day of the 6th month and continuing into the beginning of the following month, Mars moved along the northern edge of the asterism, but without actually entering it. As the term chung is not used in the text, the description may not be too critical. Here the date is probably correct.

Once more, we have a vindication of the reliability of ancient Chinese astronomical observations. It does seem that where only the month is recorded, errors of a month in the actual date are quite frequent, but this is the normal limit. It is significant that the one example of a major error occurs in (b), for which there is marked discord between the two historical sources.

In view of the above remarks, we should regard the agreement in date between the two separate records of each star as confirmation of reliability. However, whereas an error of a month in the first or last sighting of the AD 393 object would not alter the fact that the duration was unusually long, it would seriously affect the length of visibility of the AD 386 star (reported to be 3 months). The earlier guest star is thus of very uncertain nature, but the likelihood of the later star representing a supernova outburst seems fairly high.

Let us now consider what can be deduced about the position and brightness of each object in turn :

AD 386

Nan-tou, in or near which the guest star appeared, has already been discussed in Chapter 5 under the Venus observations. It is shown in Fig. 6.1 for epoch AD 386. The position of the new star in relation to Nan-tou is quite vague, for yu ("at" or "in the vicinity of") is a very general expression. It is interesting to note that the star appeared near one of the two points where the galactic equator cuts the ecliptic. None of the planets was in the vicinity while the star was visible, but it is a pity that there are no references to the Moon passing close to the star, because this might have helped to fix its position better. In AD 1605, Venus approached to within half a degree of Kepler's supernova, and this was duly recorded by the Korean astronomers (see Chapter 11). In the present case we must consider any young SNR near Nan-tou as a possible remnant. In fact there are only three remnants in the vicinity which could possibly correspond, and these are discussed below.

The brightness of the star is not estimated in the record, but we can make some useful inferences. The star was almost in opposition to the Sun throughout the period of visibility so its disappearance was not caused by heliacal setting. From the discussion in Chapter 3 of the almost complete lack of Far Eastern sightings of Mira Ceti, an apparent magnitude of +1.5 or brighter seems needed to make detection probable. Atmospheric absorption can be neglected since the meridian altitude (about 35 degrees) was fairly high. A supernova of either Type I or Type II having this apparent brightness at maximum would remain visible to the unaided eye for about 3 months. If the recorded duration is correct, a reasonable estimate of the magnitude would be near +1.5 (with some uncertainty). Following the usual practice, a distance of between 5 and 10 kpc seems probable, but the combination of poor positioning and uncertain duration make this star a rather weak supernova candidate.

From Table 4.2 only the remnant G11.2-0.3, lying a few degrees west of Nan-tou for epoch AD 386, has a distance estimate ($\geqslant 5$ kpc) within the above range. The H I absorption distance estimate corresponds to a linear diameter of the source of $\geqslant 6$ pc; the lower limit value would be unacceptably small, and even if the source were at 10 kpc (with diameter $\simeq$ 12pc),

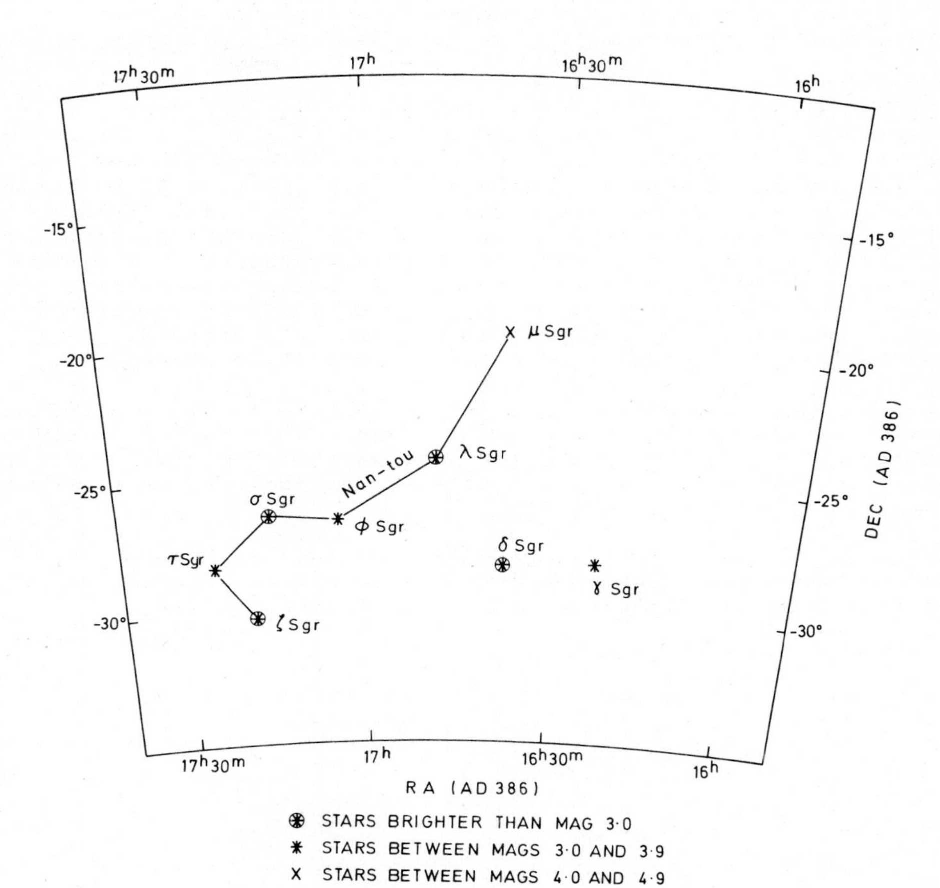

Fig. 6.1. The asterism Nan-tou.

it must have undergone a comparatively slow expansion if it was in fact the remnant of AD 386.

Only two other remnants fit the vague positional description of the guest star; these are G11.4-0.1 and G12.0-0.1, both recently discovered by Clark, Caswell, and Green (1973, 1975). Each remnant is of small angular diameter and low surface brightness, suggestive of extreme distance; in fact the Σ - D distance estimates are 15.9 and 19.7 kpc respectively. Despite the possibility that one of them may be of anomalously low surface brightness, and at closer distance, neither stands out as the likely remnant of the AD 386 event.

Because of the short recorded duration for the guest star of AD 386 and the absence of a completely acceptable remnant, this event can at best merely be described as a possible supernova.

AD 393

Here we have a much more satisfactory object. The duration is considerably longer and the position rather carefully described. The star was said to be "within" (chung) the asterism Wei. We have already discussed the use of chung (Chapter 5) and found it to be very precise. Additionally, Wei, the "Tail of the Dragon" in oriental astronomy and the "Tail of the Scorpion" in the west, is a very well defined asterism. Fig. 6.2 is drawn for the epoch AD 393, precession and proper motion having been allowed for. All stars within the boundary of the figure having an apparent magnitude brighter than +5 are represented. It requires little imagination to trace the outline of Wei.

Any star described as "within" Wei would have to lie inside the polygon completed by joining ν and ε Sco. This region is almost bisected by the galactic equator near which the majority of SNRs lie. It is only a pity that the area enclosed is rather large - some 40 square degrees.

Once more we have no direct reference to the brightness of the star. The recorded period of visibility was close to 8 months, but this does not allow for the possibility of an error of a

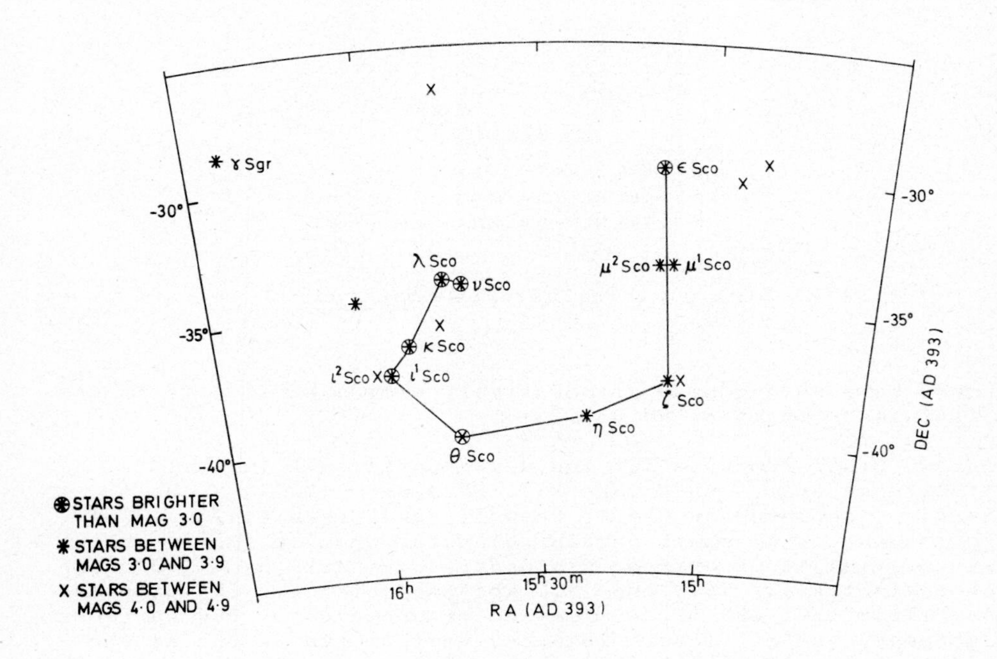

Fig. 6.2. The asterism Wei.

month in the date of first or last sighting. As it happens, the recorded date of last visibility is almost certainly in error. Even as early as the 1st day of the 9th lunar month (October 22) the star was already setting in a light sky only about 15 minutes after the Sun (from the latitude of Nan-ching, 32° N). To be still visible under such conditions would require a bright object, so that 8 months before, when the star first appeared, it would be very brilliant. Yet there is no mention in the text of recovery of the star after conjunction with the Sun (heliacal rising) or of unusual brightness. On the other hand, near the beginning of the 8th lunar month the star was setting well after sunset in a dark sky, and under these conditions quite a faint object would be discernible.

In view of the fairly frequent incidence of errors of a month in the planetary records, it would seem most reasonable to assume such is the case here rather than draw any wild conclusions about the extreme brightness of the star based on the unquestioning acceptance of the duration. This implies that we are taking the period of visibility as 7 months rather than 8, although, of course, it is quite possible that the month of discovery is in error too.

Allowing for atmospheric absorption (the altitude at Nan-ching was never more than about 20 deg.), the estimated magnitude on the assumption that the object was a supernova of Type I is close to 0 at maximum. This is similar to Arcturus or Capella or Vega. Wei lies close to the direction of the galactic centre. This, coupled with the low galactic latitude of the star (the various possible remnants lie close to the galactic equator - see below), makes it probable that interstellar absorption of light would be particularly high. A distance of no more than about 5 to 6 kpc is thus suggested. "Within" (chung) the bowl of Wei, the catalogue of Clark and Caswell (1976) lists seven SNRs: G344.7-0.1, G346.6-0.2, G350.0-1.8, G350.1-0.3, G348.5+0.1, G348.7+0.3, and G349.7+0.2. The positions of these seven SNRs are shown in Fig. 6.3 along with the outline of the asterism Wei. The epoch is 1950.

On the basis of their being of low surface-brightness and small angular diameter, and therefore probably at large distance, G344.7-0.1, G346.6-0.2, G349.7+0.2, and G350.1-0.3 might initially be eliminated from the list of candidate remnants; their Σ - D distance estimates are 15.6, 13.1, 17.4, and 18.3 kpc respectively. G349.7+0.2 also has an H I absorption result confirming a distance $\geqslant$ 10 kpc. Again there remains the possibility that one or more of G344.7-0.1, G346.6-0.2, and G350.1-0.3 may be of anomalously low surface brightness, and therefore closer than the Σ - D estimates suggest - in the absence of independent distance estimates there is no way of checking for this possibility. However, even if closer than 10 kpc, on the basis of their all having small angular diameter alone none would appear to be a likely remnant for the AD 393 event.

We will now consider the three remaining candidate remnants for AD 393 in turn.

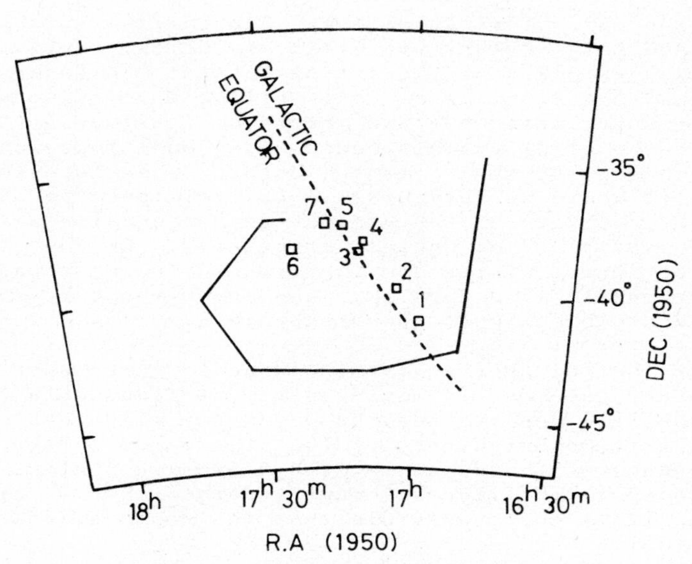

Fig. 6.3.. Catalogued SNRs within Wei.
1. G344.7-0.1 2. G346.6-0.2
3. G348.5+0.1 4. G348.7+0.3
5. G349.7+0.2 6. G350.0-1.8
7. G350.1-0.3

The source G350.0-1.8 appears as a large angular diameter (> 40 arc min) low surface brightness feature, which might be only part of a much larger remnant. The map of the source from Clark, Caswell, and Green (1975) is shown as Fig. 6.4. The Σ - D distance is 4.3 kpc and the linear diameter >36 pc. The low surface brightness and large linear diameter are suggestive of an "old" remnant (> 10,000 years), making G350.0-1.8 at best a rather weak candidate. The expected diameter of the remnant of AD 393 would be ≃15 - 20 pc; G350.0-1.8 could only have a diameter of this order if at a distance of about 2 kpc, in contrast to the distance estimate for the AD 393 supernova of up to 5 - 6 kpc. A distance as low as 2 kpc is very unlikely, since a supernova this close to the Earth would be expected to be of extreme brilliance and of long duration (as for example the supernova of AD 185).

The two SNRs G348.5+0.1 and G348.7+0.3 are shown in Fig. 6.5; this is the 408 MHz Molonglo map of Clark, Green, and Caswell (1975). It has been suggested that the two remnants may be associated because of a "bridge" linking them on high-frequency maps (for example Milne, 1969; Dickel et al, 1973). The "bridge" is also evident on the 408 MHz Molonglo map, but is not as prominent, suggesting that it has a flatter spectrum than the two non-thermal sources; it may merely be the result of confusion with an H II region along the line-of-sight. We will consider the two sources to be independent SNRs, and will place no

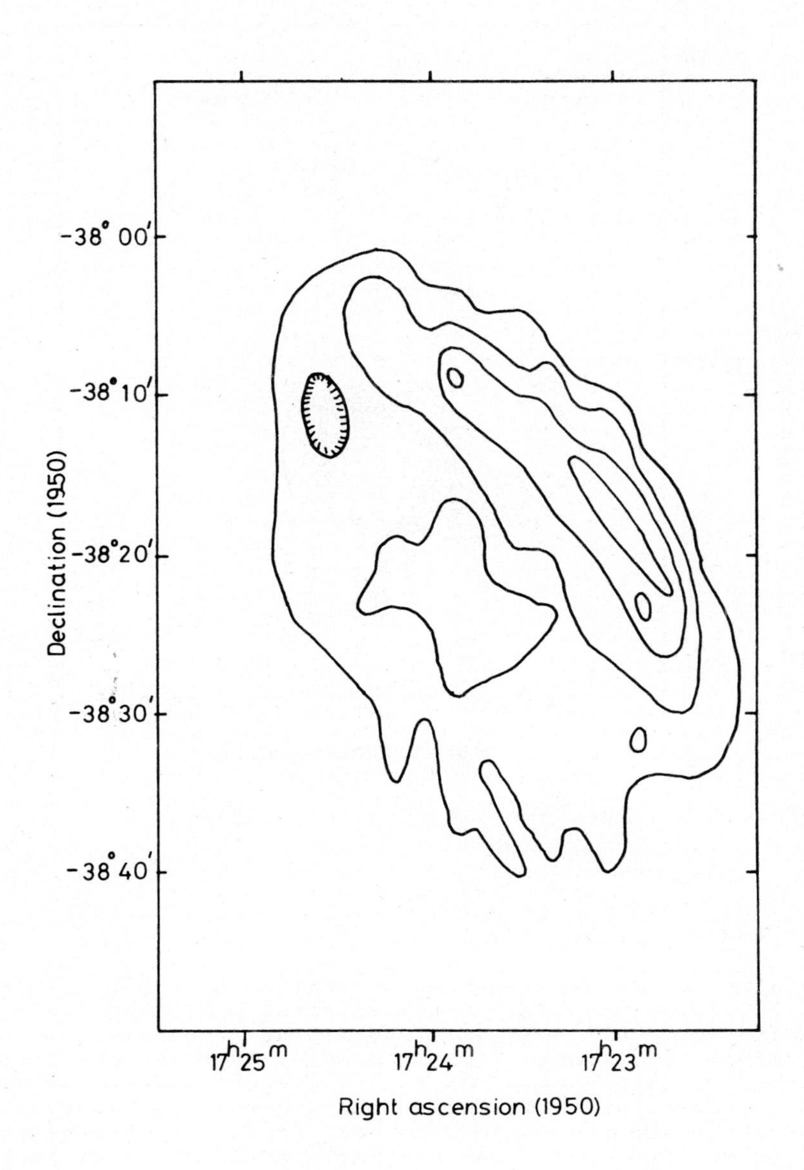

Fig. 6.4. The SNR G350.0-1.8.

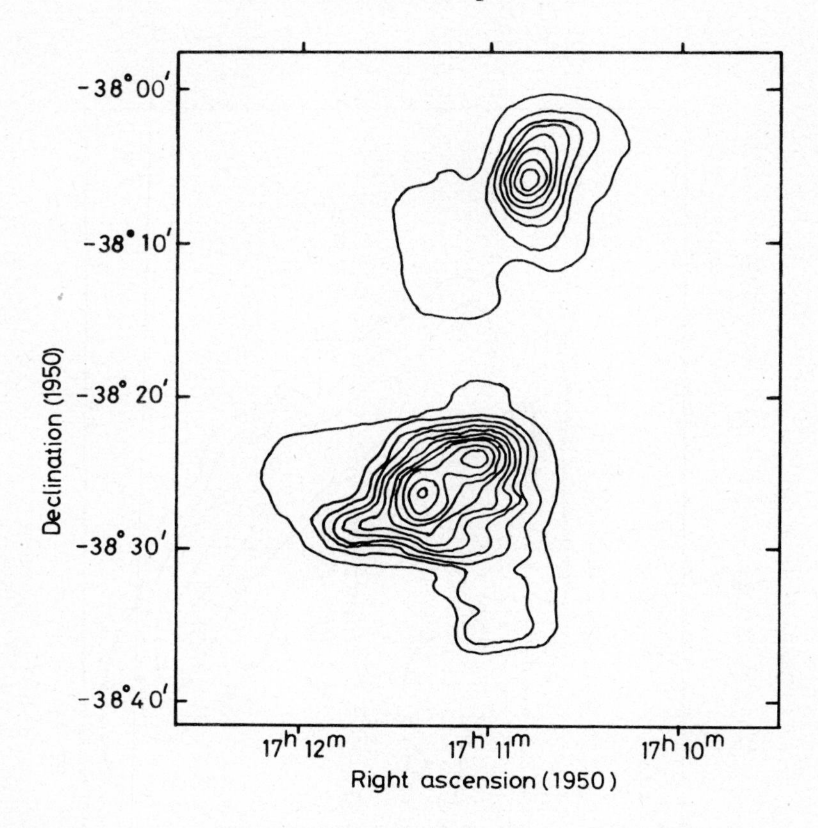

Fig. 6.5. The SNRs G348.5+0.1 and G348.7+0.3.

consequence on their positional proximity, similar surface brightness and spectral index.

The H I absorption measurements of Caswell et al (1976) place both G348.5+0.1 and G348.7+0.3 at a distance $\geqslant$ 6.7 kpc, with corresponding diameters of $\geqslant$16 pc and $\geqslant$10 pc respectively. The lower limit is in fact slightly greater than the preferred distance for the supernova of up to 5 - 6 kpc; however in view of the many uncertainties involved, the discrepancy is of no real consequence and the above data are compatible with either source being the remnant of the supernova of AD 393. It is unlikely that any other candidates for the remnant of this supernova will be found in the future, since the region has now been well surveyed.

Although there is ample vindication for the essential reliability of the astronomical records of the Chin Dynasty, the brevity of the descriptions of the two "long" duration guest stars of AD 386 and 393 makes an interpretation of their exact nature and behaviour almost impossible. The evidence for AD 386 being a supernova remains, at best, circumstantial; the uncertain duration and brightness of the event leaves open a nova interpretation, and even if it was a supernova the vague positional description makes an SNR association difficult. The long period of visibility of the AD 393 guest star means it was almost certainly a supernova. While the positional description is essentially an exact one, the size of the asterism Wei belies any attempt at an unambiguous SNR identification; nevertheless, the scope for possible equivalences is small.

Six centuries were now to pass before another supernova was seen from Earth. It was to be the most brilliant ever witnessed.

Chapter 7

AN EXTENSIVELY OBSERVED NEW STAR OF EXTREME BRIGHTNESS

After AD 393, there is a gap of more than 6 centuries, during which no supernova is known to have been observed. The star which blazed forth in AD 1006 was unquestionably a supernova, and judging from the numerous descriptions from various parts of the world, it was by far the brightest new star on record.

As discussed in Chapter 2, the intellectual climate in Europe which prevailed before the Renaissance was anything but favourably disposed towards the recording of a new star, and the Arabs seem to have been scarcely more concerned with such matters. However, because of its extreme brilliance, and despite its southerly declination, the supernova of AD 1006 aroused a fair amount of interest in the West, while in both China and Japan it was very carefully observed. Indeed, not until AD 1572 do we find another new star which is so well documented.

That we today know so much about the supernova is largely due to the patient work of Kanda Shigeru (1935) and Goldstein, the latter both independently (1965) and in collaboration with Ho Peng Yoke (1965). This latter paper by Goldstein and Ho will be referred to as GH in the rest of this chapter.

In his monumental catalogue of astronomical records of all kinds recorded in Japanese history (often in obscure sources), Kanda published a number of independent accounts of the star in his section on comets and novae. Goldstein brought to light several Arabic and European records of the new star and corrected errors made by previous investigators. Ho Peng Yoke in GH uncovered a number of Chinese observations and translated the Japanese descriptions quoted by Kanda.

We have consulted all of the original Chinese and European records of the star cited by Goldstein and GH and the more important Japanese records, and retranslated all the important sections. Some of the Japanese records are only of academic interest since they tell us almost nothing about the star. In such cases we have taken the translations direct from GH. Neither of us knows any Arabic. Copies of the Arabic texts which Goldstein used are provided at the end of his paper. We are grateful to Dr. M. Y. Tamar-Agha of the University of Newcastle upon Tyne for translating these for us. Only where there is major discord between our own translations and those of GH or Goldstein shall we make comment. It should be mentioned here that in rendering Japanese and Korean accounts in this and succeeding chapters we have preferred to give the Chinese names for asterisms and the days of the sexagesimal cycle. We feel that this should help to avoid possible confusion, and in any case the Chinese characters are, of course, used in the original texts.

The three supernova candidates discussed in previous chapters were seen only in China. In considering the new star of AD 1006, it thus seems appropriate to begin with the Chinese records. These will be followed in turn by reports from Japan, Korea, the Arab territories and Europe.

1. China

(a) Sung-shih astronomical treatise (chapter 56). "3rd year of the Ching-tê reign period, 4th month, (day) wu-yin. A chou-po ("Earl of Chou") star was seen. It appeared to the south of Ti and 1 deg. west of Ch'i-kuan. Its form was like the half Moon, with pointed rays shining so brightly (huang-huang) that one could see things clearly (chien). It passed through (li) the east of K'u-lou, and in the 8th month, following the wheel of heaven (t'ien-lu) it entered the horizon (cho). In the 11th month it was again seen at Ti. From this time onwards it regularly appeared at the hour of ch'en (7 - 9 a.m.) during the 11th month at the east, and during the 8th month at the south-west it entered the horizon".

The observations in the astronomical treatise of the Sung-shih are classified in a variety of sections. We do not find the record of the new star under the section entitled k'o-hsing ("guest stars"), but it is included in a separate short list of ching-hsing ("auspicious stars"). The reason for this will appear in due course. At certain points in the text this account is particularly difficult to translate. We have followed GH in rendering cho ("the turbid one" or "the turbid regions") as "the horizon". It would be difficult to suggest a viable alternative.

In the above record, the discovery date corresponds to AD 1006 May 6. For this same year the month of heliacal setting was equivalent to August 27 to September 24, while the star was sighted again after conjunction with the Sun between November 24 and December 22. It is clear from the above account that the star was seen for nearly two years at the very least, i.e. there were at least two heliacal settings and risings. However, the text would seem to imply a visibility for several years. This lengthy duration immediately precludes the possibility of the star being a comet (see Chapter 8) so that we are left with only a nova or supernova. The use of the term li ("passed through") in the text would seem to imply motion. However, there is ample evidence from other sources that the star remained fixed, and apart from the lengthy duration this is also implied by the statement a little later in the same text that the star reappeared at Ti. GH merely translate li as "appeared at".

(b) Sung-shih annals (chapter 7). "(3rd year of the Ching-tê reign period) 5th month, (day) jen-yin. A chou-po star was seen" - "11th month (day) jen-yin. The chou-po star was again seen".

The above dates correspond to AD 1006 May 30 and November 26. It seems probable that the second entry is giving us the true date of heliacal rising of the star, since the date is at the very beginning of the month when according to the *Sung-shih* astronomical treatise the new star reappeared. Otherwise, the annals tell us nothing. Referring to Chapter 2, we have here an example of the gross disparity between the technically accurate statement found in the *chih* and the vague notices in the *pên-chi*.

(c) *Sung-shih biographies*. "During the 3rd year of the Ching-tê reign period a large star appeared at the west of *Ti*. No-one could determine (its significance). Some said it was a *kuo-huang* or 'baleful star' (*yao-hsing*), which portended warfare and ill-fortune. At that time (Chou) K'o-ming was away on a mission to Ling-nan. On his return he urgently requested permission to reply (to these suggestions). He said, 'I have checked the *T'ien-wên-lu* and the *Ching-chou-chan*. The interpretation is that the star should be called a *chou-po* star, which is yellow in colour and resplendent in its light (*huang-huang-jan*). The country where it is visible will prosper greatly, for it is an auspicious star (*ching-hsing*). On my way back I heard that people inside and outside the court were quite disturbed about it. I humbly suggest that the civil and military officials be permitted to celebrate in order to set the Empire's mind at rest. The Emperor approved and acceded to his request. He then promoted him to the post of Librarian and Escort of the Crown Prince .....".

This fascinating account is to be found in the biography of Chou K'o-ming (AD 954 - 1017). More than half of the biography is devoted to his grandfather Chou Chien, who was a great astrologer. By his ingenuity, Chou K'o-ming averted what might have been a calamity, and evidently deserved his promotion. The above account is particularly interesting in that it explains why the star was listed as a *chou-po* rather than as a *k'o-hsing* in the astronomical treatise, and at the same time indicates that the star had attracted widespread attention.

(d) *Sung-shih biographies*. "When the *chou-po* star appeared, the star-clerk reported that it was an auspicious omen, and all the officials bowed to congratulate the emperor. Chang Chih-po gave his opinion, saying that the ruler should consolidate his virtues in response to the signs seen in the heavens, so that the star, which was not bound by any law, would remain visible. He then explained the necessity of maintaining a good government. The emperor told his officials that Chang Chih-po had the welfare of his empire at heart".

The source of this account is the biography of Chang Chih-po. The precise date of the events in unknown. GH point out that another *chou-po* is reported in the *Sung-shih* annals 10 years later and that Chang Chih-po might be referring to this. However, the latter star is of doubtful reliability since it is not reported elsewhere. Possibly it represents no more than a misplaced sighting of the AD 1006 star.

(e) Yü-hu-ch'ing-hua. "During the 3rd year of the Ching-tê reign period, a huge (t'ai) star was seen in the sky at the west of Ti. Its bright rays were like a golden disc. No-one could determine (its significance). Chou K'o-ming, the Chief Official of the Spring Agency, said that according to the T'ien (-wên)-lu and the Ching-chou-chan the star was a chou-po. It is said that its colour is golden and its rays resplendent. Its appearance brings peace and prosperity to the state concerned". The Yü-hu-ch'ing-hua was written by the Buddhist monk Wêng Ying about the year AD 1078. The above account is essentially a summary of (c). It uses much the same phraseology and therefore cannot be independent of it. However, the reason for this is obscure.

(f) Ch'ing-li-kuo-chao-hui-yao. "3rd year of the Ching-tê reign period, 5th month, 1st day. The Director of the Astronomical Bureau reported that at the first watch of the night, on the 2nd day of the 4th month, a large star, yellow in colour, appeared to the east of K'u-lou at the west of Ch'i-kuan. Its brightness had gradually increased. It was found in the 3rd degree east of Ti. Hence it belongs to the division of Chêng and the station of Shou-hsing. The star later increased in brightness. According to the star manuals there are four types of auspicious stars (ching-hsing). One of these is called a chou-po; it is yellow and resplendent and forebodes great prosperity to the state over which it appears".

According to GH, the Ch'ing-li-kuo .... was a book presented to the Sung emperor in AD 1044 by Chu Shu and his collaborators. The work dealt with court matters in the period AD 960 - 1043. The passage on the guest star exists in a preserved fragment of the Ch'ing-li-kuo ... in a modern work, the Sung-hui-yao-kao written by Ch'en-yüan (1935). The above description gives two details not found in any other work - the gradual increase in brightness of the star and the right-ascension (the third degree east of Ti). The date of the Astronomer-Royal's report corresponds to AD 1006 May 30, and the date of discovery of the star in China to May 1.

GH point out that the interpretation of a chou-po star was very subjective. In two separate places in the astronomical treatise of the Sui-shu (chapter 20) the following descriptions of the star may be found:
(i) It has a brilliant yellow colour and brings prosperity to the state over which it appears".
(ii) It is large and is of a brilliant yellow colour. Over the state it appears it presages military action, death and country-wide famine so that the population must seek refuge from their homes" (trans. GH).

It would appear that the precise interpretations depended very much on circumstances and on the sagacity of the individual. It is doubtful whether any significance should be placed on the colorimetry, since yellow was the imperial colour of the Sung Dynasty. (Each dynasty adopted one of the five colours blue, red, yellow, white, black in rotation as the imperial colour).

2. Japan

(g) Meigetsuki (volume 3). "3rd year of the Kankō reign period of Ichijō In, 4th month, 2nd day, kuei-yu. After nightfall, within (chung) Ch'i-kuan there was a large guest star. It was like Mars, and it was bright and scintillating. It was seen clearly for successive nights in the south. Some suggested that it might be due to a structural change in Ch'i-chên-chiang-chün itself".

As discussed in Chapter 3, the Meigetsuki was the personal diary of Fujiwara Sadaie, a 13th century poet-courtier, who had a special interest in guest stars. The date given by Fujiwara corresponds to AD 1006 May 1, precisely the same day that the star was discovered in China - see (f) above. Several of Fujiwara's entries concerning new stars are very detailed, and there is no doubt that he had access to a variety of original sources, some of them no longer extant. However, he never specifies his source.

(h) Ichidai yoki ("3rd year of the Kankō reign period) 3rd month, (day) wu-tzŭ amended to kêng-wu in various commentaries. A guest star entered Ch'i (-kuan). Its colour was white-blue. An instructor in astronomy, Abe Yoshimasa, reported this".

The Ichidai yoki is a Japanese record of important court events. Its authorship and date are unknown. There was no wu-tzŭ day in the 3rd month, and the amended kêng-wu corresponds to AD 1006 April 28. This is three days earlier than the date in (f) and (g), but the last day of the 3rd month was April 29 so that this amendment may be correct.

The following translations of relatively unimportant Japanese records are taken direct from GH, with only slight amendments (as mentioned above, Chinese names are used for asterisms and days of the sexagesimal cycle). Following GH, unnecessary repetition of the year (the 3rd year of the Kankō reign period) is avoided - unless of course a different year is given. Although of little astronomical value, these accounts give us a fascinating insight into the general reaction of the people at the appearance of the star.

(i) Gyokuyo (chapter 3). "The tenth day of the fourth month in the first year of the Kao reign-period (8 May 1169). Clear sky. Abe Yasuchika, the Assistant Astronomer, came in the evening ... The following was said in conversation: During the time of (emperor) Dai Sanjō In (reign: 1011 - 1016) there was an argument between Ariyuki and Morohira. Ariyuki said that what appeared within (the neighbourhood of) Ch'i-chên-chiang-chün was a 'guest star', (but) Morohira said that it was not a 'guest star'. Rather, he attributed this to the complete appearance of all the stars in (the asterism) Ch'i-chên-chiang-chün. Morohira found more support in their discussion and Ariyuki was much enraged. It was prognosticated that an important national event would take place within the next three days, and during that time (the emperor) Sanjō In had cause for anxiety. Hence what Morohira said was regarded as the correct explanation .....".

The Gyokuyo is a diary written by Fujiwara Kanezane and covers the period AD 1164 - 1200.

(j) Gonki. The twenty-fourth day in the sixth month, a chia-wu day (21 July 1006) .... a report on the 'guest star' was read (before the emperor). The twenty-fifth day, an i-wei day (22 July) ... reports from the various routes on the guest star were read. The third day of the seventh month, a kuei-mao day (30 July) ... from behind the curtain (which separated the emperor from his subjects) the Second Minister gave reports from the various routes, one on the magical mirror and one on the 'guest star' - the matter concerning the 'guest star' was still undecided. The thirteenth day in the seventh month, a kuei-ch'ou day (8 August) ... (the author) went to (the palace); a decision was made on the '(guest) star' ".

The Gonki is a diary written by Fujiwarano Yukinari (AD 971 - 1027), and covers the period AD 991 - 1011.

(k) Hyakurensho. "During the fourth month (May 1006) a large star appeared in the SE. On the thirteenth day of the seventh month (9 August 1006) the officials made a decision on the reports coming from the various routes about the 'guest star' ".

The Hyakurensho is a history of the imperial court from AD 967 to 1259. It was written around AD 1260, but the authorship is unknown.

(l) Hōjōji Sesseiki. "The thirteenth day in the seventh month, a kuei-chou day (9 August 1006) ... a decision was made regarding the reports from the various routes on the large 'guest star' ... The nineteenth day, a chi-wei day (15 August) ... the cabinet ordered that divination be made regarding the large 'star'.... The eighth day of the eighth month, a wu-yin day (3 September) .... offerings were made to twenty-one shrines .... The twenty-sixth day (21 September), a ping-shên day .... the Second General came and asked for an amnesty because of the appearance of the large star".

(m) Nihonkiryaku. "On the thirteenth day of the seventh month, a kuei-ch'ou day (9 August 1006) the lords and ministers decided on the reports from the various routes regarding the 'guest star'. On the nineteenth day, a chi-wei day (21 August) the soothsayer made divination because of the guest star. On the nineteenth day in the eighth month, a chi-ch'ou day (14 September) offerings were made to the various shrines because of the guest star".

From the above records, it is evident that the new star was still attracting general attention as late as mid-September - $4\frac{1}{2}$ months after discovery.

3. Korea

(n) Koryŏ-sa (chapter 47). "During the 9th year of Mokchong a hui-hsing ('broom star') was seen".

This is the only astronomical record from Korea for the year AD 1006. It is uncertain whether this has anything to do with the new star reported in China and Japan. Although there is no mention of a comet being observed in these countries in AD 1006, a comet is reported in the astronomical treatise of the Sung-shih in the previous year. It is not until AD 1011 that regular astronomical records commence in Korea. Before then, an error of a year in the recorded date of an event appears quite possible. It seems best to leave the question of this Korean sighting open.

4. The Arab Dominions

(o) Alī ibn Ridwān : Commentary on the Tetrabiblos of Ptolemy. "I will now describe a spectacle (athar) which I saw at the beginning of my studies. This spectacle (athar) appeared in the zodiacal sign Scorpio, in opposition to the Sun. The Sun on that day was 15 degrees in Taurus and the spectacle (nayzak) in the 15th degree of Scorpio. This spectacle (nayzak) was a large circular body, $2\frac{1}{2}$ to 3 times as large as Venus. The sky was shining because of its light. The intensity of its light was a little more than a quarter of that of moonlight . It remained where it was and it moved daily with its zodiacal sign until the Sun was in sextile with it (i.e. 60 deg. away) in Virgo, when it disappeared at once (duf'atan wāhidatan).

All I have mentioned is my own personal experience, and other scholars from my time have followed it and came to a similar conclusion. The positions of the planets at the beginning of its appearance were like this: the Sun and Moon met in the 15th degree of Taurus; Saturn was 12° 11' in Leo; Jupiter was 11° 21' in Cancer; Mars was 21° 19' in Scorpio; Venus was 12° 28' in Gemini; Mercury was 5° 11' in Taurus; and the Moon's node was 23° 28' in Sagittarius. The spectacle (nayzak) occurred in the 15th degree of Scorpio. The ascendant of the conjunction when the spectacle (nayzak) appeared over Fustat of Egypt was 4° 2' in Leo. Also the tenth (house which included most of) Taurus began at 26° 27' in Aries.

Because the zodiacal sign Scorpio is a bad omen for the Islamic religion, they bitterly fought each other in great wars and many of their great countries were destroyed. Also many incidents happened to the king of the two holy cities (Mecca and Medina). Drought, increase of prices and famine occurred, and countless thousands died by the sword as well as from famine and pestilence. At the time when the spectacle (nayzak) appeared calamity and destruction occurred which lasted for many years afterwards ....".

Goldstein located the above text in ms. Escurial (Casiri) - he gives a full reference. He also points out that a Latin translation of the passage may be found in Aegidius Tebaldinus, Quadripartitum (Venice, AD 1493). Quadripartitum is the Latin equivalent of the Greek Tetrabiblos. This latter manuscript gives the longitude of Mars as 21° 9' in Scorpio, and of Mercury as 5° 5' in Taurus. Additionally the beginning of the tenth house is given as 27° 27' in Aries. In rendering the above text, we have followed Goldstein in translating both athar and nayzak as "spectacle". He was here being cautious in order to avoid prejudicing the interpretation of the phenomenon.

We shall discuss later the date of the phenomenon as determined from the astronomical data, but Goldstein points out that a date of occurrence in AD 1006 would fit in well with the youth of Alī ibn Ridwān (he died in AD 1061).

(p) Ibn al-Athīr: (A.H. 396). At the beginning of Sha'ban, a large star similar to Venus appeared to the left of the qibla of Iraq. Its rays on the Earth were like the rays of the Moon and it stayed until the middle of Dhū al-Qa'da and disappeared".

Ibn al-Athīr lived during the 13th century of our era. Goldstein showed that the above account is undoubtedly taken from the description by Ibn al-Jawzī (died AD 1200) in his book Kitāb al-Muntazam. This is as follows :

(q) "Year 396. Among the incidents in that year a large star similar to Venus in size and brightness glittered to the left of the qibla. Its rays on the Earth were like the rays of the Moon. This was on the night (preceding) Friday, the beginning of Sha'ban, and it stayed until the middle of Dhū al-Qa'da and disappeared".

The date of appearance of the star corresponds to AD 1006 May 3, and the date of disappearance to around August 13 in the same year.

(r) Bar Hebraus: "A.H. 396: There appeared a great star resembling Aphrodite in greatness and splendour in the zodiacal sign Scorpio, its rays revolved and gave out light like that of the Moon - it remained four months and disappeared".

The above translation from the Syriac is taken direct from Goldstein, who acknowledges Budge (1932) as his source. Bar Hebraus lived a little later than Ibn al-Athīr, and would appear to be abstracting from him or Ibn al-Jawzī. At any rate, he adds nothing new.

(s) Annales Regum Mauritaniae "In the year 396 (A.H.) a great (variant: strange) tailed star appeared, bright (variant: burning). (Rather long concentration was needed to see its movement). It was one nayzak of the twelve, which was mentioned by the ancients, and their scholars studied it over a long period. They claimed that this star appears to a certain purpose which God plans in the world and only God knows. (It appeared on the first of Sha'ban, 96 mentioned above {i.e. 396 A.H.}). Its first appearance was before sunset, then it faded until night came and it appeared again. This star stayed for six months. During this year it was stormy and windy, and there was thunder without rain".

The Annales Regum Mauritaniae is an Arabic chronicle extending to A.H. 726. Precisely the same date of first appearance is given as in (p) and (q) above. The observation could have been made in north-west Africa or Spain.

It would not be proper to leave this section on the Arabic observations of the new star of AD 1006 without making reference to the spurious supernova of AD 827. Humboldt (1851) in his list of historically observed new stars included an object under the year AD 827 observed by Arabian astronomers. However, he believed that the precise year was doubtful, stating, "It may with more certainty be assigned to the first half of the ninth century, when in the reign of Caliph Al Mamoun the two famous Arabian astronomers, Haly and Giafar Ben Mohammed Albumazar observed at Babylon a new star, whose light, according to their report, 'equalled that of the Moon in her quarters'. This natural phenomenon likewise occurred in Scorpio. The star disappeared after a period of four months".

This star found its way into the catalogue of Lundmark (1921), and thence into the more recent works of Hsi Tsê-tsung (1955) and Hsi Tsê-tsung and Po Shu-jen (1965). However, from the investigation of early sources, Goldstein was able to show conclusively that "Haly" was none other than Alī ibn Ridwān, and that the true date of the star was AD 1006. Tammann (1966) had gone so far as to link the supposed supernova of AD 827 with the X-ray source Sco X-3, but in a brief note, Goldstein (1966) corrected his error. Here is ample evidence that there is no substitute for consultation of original historical sources.

5. Europe

The following references are all from the compilation of monastic chronicles Monumenta Germaniae Historica (Pertz, 1826→).

(t) Annales Beneventani. "1006. In the 25th year of our master Pandolphus and the 19th year of our master Landolphus, his son, a very brilliant (clarissima) star shone forth, and there was a great drought for three months".

The annals of the monastery of St. Sophie, Benevento, extend from the mid-8th century AD to 1128. They are much concerned with very local events, and in all probability the star was seen in Benevento.

(u) Annales Sangallenses Maiores. "1006. A new star of unusual size appeared, glittering in aspect, and dazzling (verberans) the eyes, causing alarm. In a wonderful manner this was sometimes contracted, sometimes diffused, and moreover sometimes extinguished. It was seen likewise for three months in the inmost limits of the south (in intimis finibus austri), beyond all the constellations which are seen in the sky".

The reliable section (Pars Altera) of the chronicle of the Benedictine monastery of St. Gallen covers the period from AD 919 to 1044. Like the chronicle of Benevento, it is mainly concerned with local affairs, although it makes frequent reference to major European events. There seems no reason to doubt that the observations of the star were made in St. Gallen itself. Certainly the description, which is probably that of an eyewitness, is unique. One manuscript gives the date as AD 1012, but after AD 974 this differs by 6 years from the other manuscripts for almost all dates. Fortunately we can check which series of dates is correct from observations of comets reported in AD 975 autumn (981 in the discordant manuscript), 989 Feast of St. Laurence, i.e. August 10, (995) and 998 February (1003). Oriental records in the catalogue of Ho Peng Yoke (1962) give details of comets in AD 975 August, 989 August and 998 February. No comets are recorded in AD 981 and 995, and the comet of 1003 was seen in December. It is clear that the new star which interests us was seen in AD 1006 in St. Gallen. The record of the star is the only entry for the year, emphasising the importance attached to it - comparable to a famine in the previous year and a plague in the following year.

(v) The *Annales Laubienis* and the related *Annales Leodiensis* both mention a comet in the year AD 1006. The records run as follows: "1006. There was a very great famine and a comet appeared for a long time."

At this period both chronicles are fairly general, covering events throughout Europe. Presuming that the "comet" refers to the new star, as seems likely from its long duration, there seems no reason to believe that it was seen as far north as Belgium. In a valuable note, Porter (1974) has pointed out that the Belgian records may be derived from St. Gallen. He states that a member of the influential St. Gallen family of Notker, a nephew of Otto I, was bishop of Liege and also Abbot of Lobbes, where the two chronicles were compiled.

(w) *Iohannis Chronicon Venetum*. "And so at the same time, a comet, the sign of which always announces human shame, appeared in the southern regions, which was followed by a great pestilence throughout all the territories of Italy or Venice .....".

The *Chronicon Venetum* begins in the 6th century AD and continues to AD 1008. The latter part of the work is generally concerned with events in Venice itself, and it seems quite possible that John witnessed the star he mentions. However, the precise year is not given, and the last datable reference is June in AD 1004. His comment, "And so at the same time ....." may suggest an earlier event than AD 1006, although the positional description "in the southern regions" is essentially correct for the supernova. Our author may have been ignorant of the true nature of the star, but in any case he is mainly concerned with what is portended.

Our attention was drawn to the following two sources by Ms. H. H. Warwick of Minneapolis (both are to be found in *Monumenta Germaniae Historica*):

(x) *Alpertus de Diversitate Temporum, Lib. I*. "Three years after the king (Henry II) was raised to the throne of the kingdom, a comet with a horrible appearance was seen in the southern part of the sky, emitting flames this way and that".

Alpert of Metz (France) was a contemporary, and so his account was probably original. Henry II ascended the throne in AD 1002 following the death of Otto III, so that the date of the "comet" would correspond to AD 1005. Errors of a year are of frequent occurrence in medieval European chronicles (c.f. Newton, 1972), so here we may have another allusion to the supernova, with the "flames" being the result of atmospheric turbulence at a low altitude. A definite conclusion is not possible.

(y) Annales Mosomagenses. "1006. In this year was seen in the sky a burning star like a torch which is called a comet".

The annals of Mousson (France) appear to be original at this period The year is correct for the supernova, so that we may presume that the supernova is referred to. However, just how useful the record is can be judged from the fact that in the same annals the comets of AD 1066 (Halley's) and 1105 are described in exactly the above words.

Both of these last two accounts are copied either verbatim or in a simplified form in a wide variety of European chronicles. Apart from the St. Gallen sighting, which more or less accidentally sets a valuable southern limit to the declination, the European observations tell us virtually nothing about the star.

As far as the nature of the star is concerned, several records rule out the possibility of it being a comet. We have the very long duration (several years) in (a), while (a), (f), (o), (q), (s) and (u) imply a fixed location over several months. A bright comet moves at something like 10 deg. a day (see chapter 8) so that its position changes rapidly from night to night. We shall leave the question of whether the star was a nova or supernova until a little later.

Discovery seems to have been almost simultaneous throughout the world. In China it was first seen on a date corresponding to AD 1006 May 1 - see text (f) - and in Japan possibly as early as April 28 (h), but certainly by May 1 (g). One of the Arabic reports (p), gives a date equivalent to May 3. Another report (o) is somewhat cryptic, and deserves special discussion.

Alī ibn Ridwān gave the positions of the Sun, Moon and planets when the star first became visible. The longitude of the Moon is the most useful since the Moon moves at a mean daily rate of some 13 deg. As Goldstein pointed out, Alī ibn Ridwān's data were calculated rather than observed, as is clear from the fact that the Moon was in conjunction with the Sun at the time and thus invisible. However, this need not affect their reliability. Conjunction of the Moon with the Sun occurred on April 30, and we have interpolated the various longitudes (with the exception of the node, which we have calculated direct) from the tables of Tuckerman (1964). Tuckerman gave his figures for 16^h U.T. (roughly 6 p.m. in Egypt) and we have used this time on April 30 as a preliminary moment for calculation. Alī ibn Ridwān's data, converted to longitude in degrees and decimals, are shown for comparison - see Table 7.1.

The position of Mars is obviously rather seriously in error, but for the other planets and the Sun the agreement is excellent. AD 1006 April 30 was unquestionably the date which Alī ibn Ridwān

	Alī ibn Ridwān	Modern (computed)
Sun	45°	44.9°
Moon	45	48.9
Mercury	35.18	35.45
Venus	72.47	72.58
Mars	231.32	227.87
Jupiter	101.35	100.70
Saturn	132.18	131.97
Ascending node	263.47	263.57

Table 7.1. Comparison of longitudes of the Sun, Moon and planets according to Alī ibn Ridwān and modern computation.

intended. However, if we make the necessary correction to the Moon's longitude in order to agree with the figure quoted by him, the local time corresponds to about 11 a.m. This might have been reasonable enough if the star was above the horizon, for it was bright enough to have been visible by day - see text (s). However, the account expressly states that the star was in direct opposition to the Sun, and thus well below the horizon at the calculated time. Both the position of the "ascendant of the conjunction" (i.e. the point of the ecliptic which was rising at the time) and that of the "beginning of the tenth house" (i.e. the momentary point of upper culmination of the ecliptic) are in agreement with a time around local noon. The most likely explanation that we can offer is that the star was not seen on the night of April 29 to 30, but *was* seen on the following night. For astrological purposes, Alī ibn Ridwān had to assume a precise time of first appearance of the new star, and thus chose noon (or a time close to this) on the intermediate day.

Before discussing the recorded duration, it is necessary to deduce the location of the star as accurately as possible. Several texts indicate a duration of around three months, and it is important to decide whether or not this is the result of heliacal setting. If not, it would imply a sharp decline in the light curve after this period.

Let us examine the records which are useful in determining the position of the new star. To begin with, the following tell us nothing useful about its location: (b), (d), (j), (k), (l), (m), (n), (s), (t), (v), (w), (x) and (y). In addition (c) and (e) are almost entirely concerned with the interpretation of the star as a portent, so that the rather vague allusion to its location ("at the west of *Ti*") is of little significance. We feel that the position of the star in relation to the *qibla* (i.e. the local direction of Mecca) in (p) and (q) is valueless in view of the uncertainty in the precise place of observation and the low accuracy with which the *qibla* was probably determined (Newton, 1972) Porter (1974) has discussed this last question in some detail,

Plate 8. A view of St. Gallen and the Alpstein mountains.

and finds no alternative but to reject the observation.

The various Chinese and Japanese descriptions which are of value all involve references to asterisms or lunar mansions. This is a study on its own, and we shall thus commence with the Arabic and European observations (o) and (u). Alī ibn Ridwān gave the longitude of the star as the 15th degree of Scorpio, i.e. between longitudes 224 and 225°, in AD 1006. Just how accurate this is we cannot say. Comparison with the accuracy of the longitudes of the Sun, Moon and planets which he gives cannot be made, for these were calculated. However, as the measurement is quoted to the nearest degree, it seems reasonable to assume a probable error of a similar amount, although it would be unwise to place too much emphasis on this.

One of the most fascinating observations is that made at St. Gallen in Switzerland. The record, without any pretensions to accuracy of observation, sets valuable limits to the declination of the star, since the implication is that it was only barely above the horizon. St. Gallen lies in a valley. The terrain around the town is rather hilly (on the 100 m scale), but the southern horizon is dominated by the great peaks of the Alpstein range (maximum height 2502 m), some 20 km away. Plate 8 shows a view of St. Gallen and the Alpstein mountains, looking towards the south. We have used this photograph and a detailed relief map of the area (on a scale of 1:50,000) to draw the profile of the southern horizon as seen from St. Gallen, see Fig. 7.1.

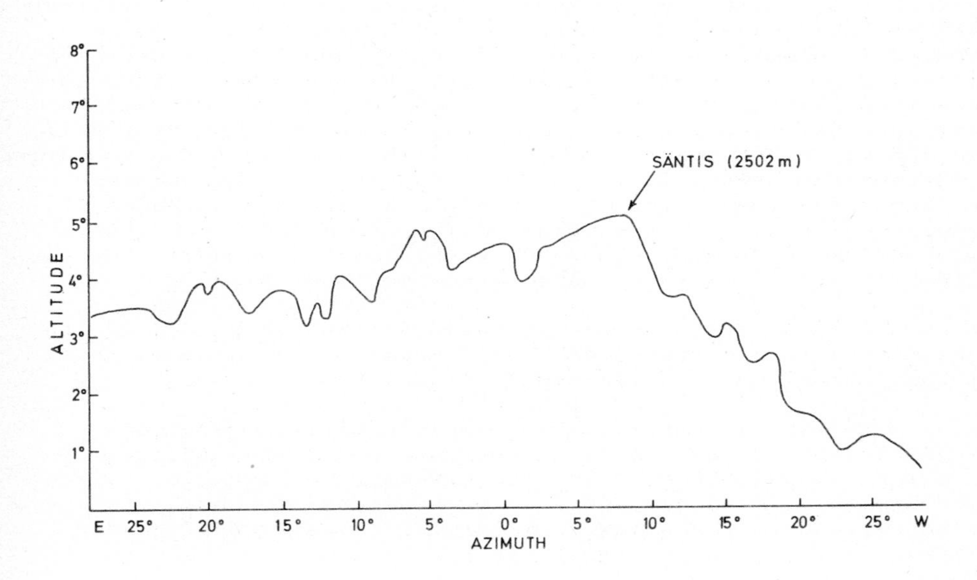

Fig. 7.1. Southern horizon profile as seen from the vicinity of St. Gallen.

The town itself is at an altitude of about 670 m above sea-level, while the surrounding hills reach a height of 800 or 900 m. In calculating the horizon profile, we have assumed a mean height of some 750 m in the vicinity of St. Gallen, but the precise height and location barely affect the result - we estimate that the profile is accurate to within a small fraction of a degree.

As the latitude of St. Gallen is 47.4°N, it is clear that, allowing for refraction, no star south of declination -38.5° would be visible from the neighbourhood of the town. On the other hand, if the star was as much as 5 deg. to the north of this, it could scarcely be described as so close to the horizon ("in the inmost limits of the south, beyond all the constellations which are seen in the sky"). When it first appeared, the star was in opposition to the Sun and was thus best placed for observation when on the meridian. We therefore set a somewhat arbitrary northern limit of declination as -33.5°, but it seems highly unlikely that the star was further north than this. The allusion to the severe flickering of the star would also suggest a very low altitude, where atmospheric turbulence is greatest.

Proceeding now to the Far Eastern records, Fig. 7.2 shows the area of the sky in which the new star appeared, with special reference to the oriental asterisms. Co-ordinates are for the epoch AD 1006 (precession has been applied, but proper motions are negligible). All of the stars within the figure which are brighter than magnitude +4.5 are represented. At the latitude of the Chinese and Japanese capitals (K'ai-fêng and Kyōto - both close to 35° N), the meridian altitude of a star in the centre of the chart would be less than 20 deg. At such a low elevation, atmospheric dimming would be significant, and stars fainter than +4.5 would not be readily discernible. On the chart, stars in the early catalogues of Bayer and Flamsteed (which were very incomplete) are denoted by their Greek letter or number as still used today; other stars are unidentified. The western boundaries of the lunar mansions Ti and K'ang are marked for reference. Also shown are the asterisms Ch'i-kuan, Ku-lou and Ch'i-chên-chiang-chün mentioned in several of the texts and also other neighbouring asterisms. In drawing this chart, we have made use of the early oriental star maps discussed in Chapter 2.

We are confident that the configurations of Chi-kuan and K'u-lou in particular are correct, and there seems to be no doubt that Chi-chên-chiang-chün is the single star κ Lup.

Fig. 7.3 is more schematic. Here individual stars are not represented. The lunar mansion boundaries and the outlines of the principal asterisms are copied direct from Fig. 7.2. Also shown are the limits deduced from the European and Arabic observations (u) and (o). Let us consider the Chinese and Japanese observations in turn, referring to Fig. 7.3.

The Chinese report (a) states that the new star appeared "to the south of Ti" , probably indicating that it was within Ti lunar mansion. Later in the same text the star is described as

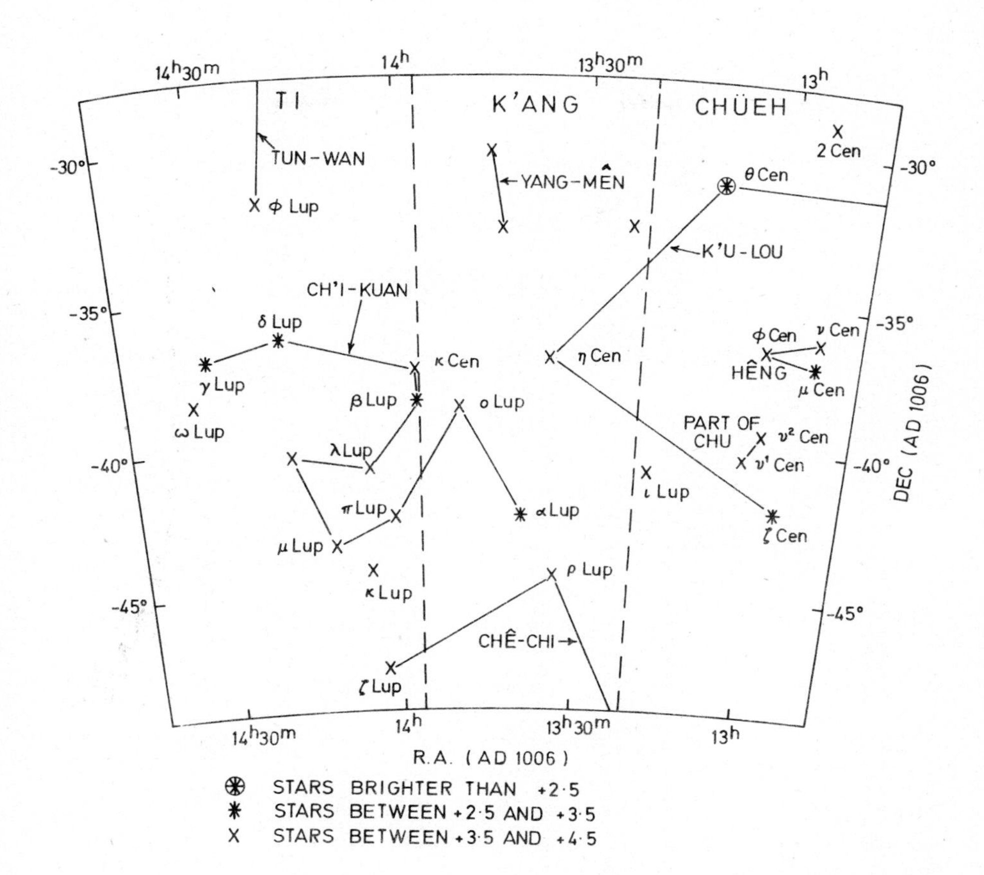

Fig. 7.2. Stars in the vicinity of the new star of AD 1006.

"at Ti", and this seems quite definite. However, if we interpret literally the statement that the star was "1 deg. west of Ch'i-kuan", this would imply a location in the western half of K'ang. This latter remark is not at all clear.

Fortunately (f) gives a precise right ascension - the 3rd degree of Ti. Coming from the Director of the Astronomical Bureau, this measurement should be reliable. Ho Peng Yoke et al (1972) quote a recent investigation by Yabuuchi (1967) of the accuracy of mid-11th century AD Chinese measurements of the positions of the determinant stars of the 28 lunar mansions. Presumably these are declinations, for the Chinese never conceived of the the idea of a single origin of R.A. (right ascensions were measured from the determinant stars themselves). The mean error found by Yabuuchi was only 0.4° with a maximum error for the sample of 1.3°. As discussed in Chapter 8, the R.A. of T'ien-kuan (ζ Tau.),

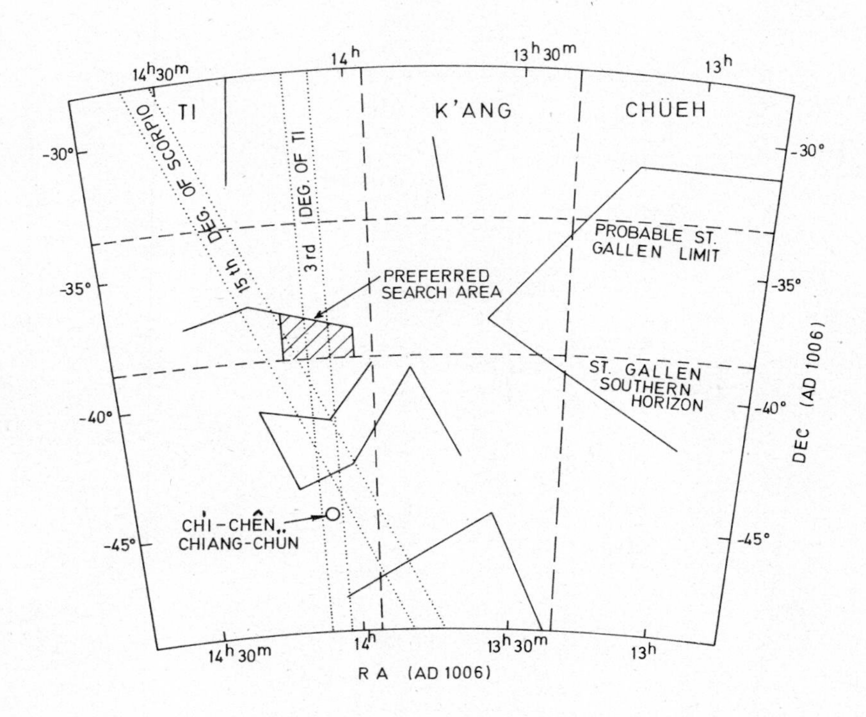

Fig. 7.3. The preferred search area for the remnant of the AD 1006 supernova.

measured at much the same time, is only about 0.9° in error. It thus seems that the uncertainty in the measured R.A. of the new star of AD 1006 should be no more than about 1 deg. Such a location would fit in reasonably well with the description "to the east of Ku-lou, at the west of Ch'i-kuan." In Fig. 7.3, the zone corresponding to the 3rd degree of Ti is shown. The agreement with the Arabic measurement (15th degree of Scorpio) is striking.

Coming now to the Japanese records, (g) describes the star as "within" (chung) Ch'i-kuan. We have already discussed (in Chapters 5 and 6) how precise the term chung is. Where Fujiwara obtained his information we cannot say, but this does not alter the fact that it is a simple matter to decide with fair precision whether or not a new star lies within an asterism. The position of the star thus indicated, which is confirmed in a more general way by text (h), will be discussed below.

The possibility of the star being in the vicinity of Chi-chên-chiang-chün (κ Lup) is ruled out by the St. Gallen sightings; κ Lup lay fully 6 deg. to the south of the St. Gallen horizon of the time. Fujiwara seems to have obtained his information on the new star from two independent sources, and GH have questioned the reliability of his reference to *Ch'i-chên-chiang-chün*. The source of Fujiwara's information here seems to be the conversation reported in the *Gyokuyo* (i). Just how accurate an account Abe Yasuchika would have of an argument which took place some 150 years before his own time is impossible to say. GH have pointed out that *Ch'i-chên-chiang-chün* is only a single star (the "Cavalry General"), whereas all the stars of the asterism are supposed to have appeared. They have suggested that *Ch'i-kuan* was intended. However, the whole statement seems to be too obscure to be of any value.

As far as the oriental records are concerned, we have so far concentrated mainly on the R.A. of the star. However, the several references to *Ch'i-kuan*, and the lack of allusion to any asterism north of it implies a northern limit close to -35° declination, in good agreement with the estimated St. Gallen northern limit.

The most probable location of the star, as we define it, is denoted by the small shaded area in Fig. 7.3. This is bounded on the east and west by the seemingly reliable one degree uncertainty in the Chinese measurement of the ("3rd degree of *Ti*"), and on the south by the firm limit set by the St. Gallen horizon. In deciding on the northern limit, we have placed a rigorous interpretation on the term *chung* in the *Meigetsuki* record (g). We feel that earlier in this book we have given ample evidence to justify this. If this particular northern limit is discounted, the declination of the star cannot be pushed more than about 3 deg. further north since the star would have then been nearer *Tun-wan* (an asterism not mentioned in any text) than *Ch'i-kuan* and at the same time well above the St. Gallen horizon.

Our preferred search area could be halved if the uncertainty in the Arabic observation (o) was limited to no more than one degree. However, there is no supporting evidence for this, and even at the extreme western edge of the shaded area an error of no more than 2 deg. is implied. We therefore estimate the co-ordinates of the star at the epoch AD 1006 as R.A. $14^h\ 10^m \pm 10^m$, dec. $-37.5^\circ \pm 1.0^\circ$ (errors are only approximate).

We are now able to discuss fully the apparent brightness of the star at maximum and its duration of visibility. It is by no means easy to form an estimate of the magnitude of the star at maximum. In Japan (g), the star was compared with Mars, then in opposition to the Sun and thus around -2. However, in Iraq (p and q), although described as "similar to Venus in size and brightness" (mag. about -4), "its rays on the Earth were like the rays of the Moon". A comparison with the Moon was also made in Egypt (o), while in China (a) and Switzerland the inference was that it was excessively brilliant.

The key to this seemingly remarkable range of brightnesses is perhaps supplied by a Chinese record (f) which twice mentions an increase in brightness. Reporting on a date corresponding to May 30 (discovery in China was on May 1), the Director of the Astronomical Bureau stated that "the brightness had gradually increased". In the same text we read, "The star later increased in brightness". This suggests an increment in brightness over several days. Presumably then, the comparisons with Mars and Venus were made in the early stages after discovery. Certainly, these two planets could never be compared with the Moon in brightness (mag. -12.5 at full, -10 at first and last quarters). We know that the star of AD 1054 (see Chapter 8) reached a peak magnitude similar to Venus, but there are no allusions to unusual brightness although we have several records from China and Japan. Yet the star of AD 1006, because of its extreme southerly declination, was in a much inferior situation.

Let us consider the various accounts which mention great brightness in order. In China, (a), the star shone so brightly (huang-huang) that one could see things clearly (chien). The use of repetition (e.g. huang-huang) is common in Chinese for emphasis. More important, the character chien, which we have rendered "see clearly in the context of the record, generally means "to scrutinise" or "to look closely at", implying a high level of illumination.

From Payne-Gaposchkin (1956), the starlight in the night sky (most of which is due to faint stars) equals the light of about 1000 first magnitude stars. However, starlight contributes only about one-sixth of the total sky brightness; the residual sky light comes from the zodiacal light and the "permanent aurora". Allowing for the fact that we can only see half the sky from any one place, and correcting for atmospheric absorption, the integrated sky brightness is equivalent to about -7.5. To produce the effect described, the star would clearly have to be significantly brighter than this.

We learn from Capt. H. Haysham, of the Marine and Technical College, South Shields, who has done much observing at sea in the Tropics, that with the Moon high in a clear sky, one can start to discern distant objects by its light when the age is about 5 days (elongation about 65 deg.). However, by the time the Moon has reached first quarter, objects can be made out clearly, and even colours begin to be distinguished. Fig. 7.4 shows the mean magnitude of the Moon as a function of elongation from the Sun (for elongations in excess of 30 deg.). Figures are based on data quoted by Fessenkov (1962). From the figure, it is evident that the 5-day old Moon reaches a magnitude close to -9, and the half Moon nearly -10. The new star of AD 1006 never attained an altitude of more than about 17.5° at K'ai-fêng, the Chinese capital of the time, so that allowing nearly a magnitude for atmospheric extinction, a result of magnitude -9.5 seems a conservative estimate.

In Egypt (o), the sky was described as "shining" because of the light of the star, and the intensity of its light was estimated as "a little more than a quarter of that of moonlight". It seems obvious that the full Moon was intended as the comparison object since otherwise we would expect the phase to be stated. Furthermore, comparison with the full Moon would be natural since the star was in opposition to the Sun and thus closest to the Moon when the latter was near full (less than 20 deg. away).

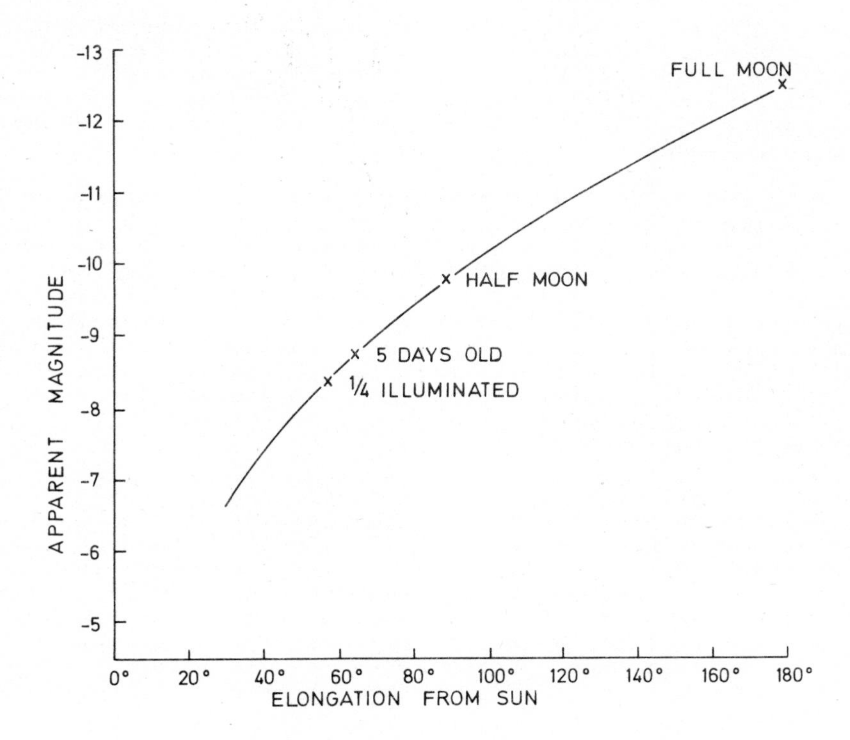

Fig. 7.4. Apparent magnitude of the Moon as a function of elongation.

From Muller and Stephenson (1975), changing light intensity has a subjective effect approximating the square root of the brightness. On this basis, the best estimate of the maximum brightness of the star would be about one-fifteenth of that of the full Moon, or roughly magnitude -9.5. This, somewhat surprisingly for such crude estimates, agrees with our previous result. According to Goldstein, a possible reading of the text is that the star was "as bright as the Moon a little more than one-quarter illuminated". The Moon at this phase has a magnitude close to -8.5. However, correcting for differential atmospheric extinction, since the Moon at first quarter was in a high northern declination, we return to -9.5 for the star.

The mention of the apparent angular size of the star in the same text ("$2\frac{1}{2}$ to 3 times as large as Venus") is difficult to interpret. Due to an optical effect whose cause is partly physiological, all stars appear to have an apparent diameter which is greater the brighter the star. Thus Tycho Brahe estimated the apparent diameters of stars of different magnitudes as follows: 1st mag. 120", 2nd 90", 3rd 65", 4th 45", 5th 30", 6th 20". Again, the mediaeval Arabian astronomer Al Fargani gave the apparent diameters of the planets as : Mercury 1/15 of the Sun, Venus 1/10, Mars 1/20, Jupiter 1/12, Saturn 1/18 (for both sets of data see Dreyer, 1906). To attempt to extrapolate on these figures in order to deduce the apparent magnitude of the new star would be mere guesswork, especially in view of the close agreement between the figures for Jupiter and Venus, but the large estimate of the apparent size of the star suggests that it was very much brighter than Venus.

From Iraq (p and q) we find a further independent comparison with the Moon, while at St. Gallen, although the light loss on account of atmospheric absorption at such a low altitude would be some 2 mag., the star could still be described as "dazzling" or "beating at" the eyes. All of the various photometric indications are necessarily very rough, but we feel that the best estimate of the apparent magnitude of the star at maximum is within 1 mag. of -9.5. (It is interesting to note that Goldstein deduced -8 to -10). A star fainter than -8.5 could scarcely produce the effects described, while if much brighter than -10.5 it would rival the full Moon.

The absolute magnitude of a nova varies from -8.5 (fast) to -6.5 (slow)- see Chapter 8. Thus if the star was a fast nova it would be at a distance of less than 10 pc from us, and if a slow nova less than 5 pc away. It would therefore be among the very nearest stars. The probability of such an event occurring during the last 2,000 years (the period of detailed recorded history) is very small. In fact, a fair estimate of the frequency of novae occurring less than 10 pc away and as close to the galactic equator as the new star of AD 1006 (some 15 deg.) is about once every million years.

Additionally, the possibility of the star being a fast nova seems to be excluded by the rather careful Chinese description (f) of its gradual increase in brightness. Fast novae tend to rise to a maximum within a day or so and then decline rapidly (c.f. Payne-Gaposchkin, 1957). The light curve of a slow nova follows no definite form, but the remnant, a white dwarf in a binary system, is typically less than 10 mag. fainter than the maximum of the outburst. In the event of the star being a slow nova, the remnant would be close to mag. 0. The brightest star shown in Fig. 7.2 is θ Cen. (mag. +2.3). All the stars represented in the figure are at a distance of more than about 50 pc with the exception of θ Cen. and 2 Cen., which are both about 20 pc away (c.f. the catalogue of Schlesinger, 1940). However, both of these stars are isolated red giants, and in any case lie two lunar mansions to the west of Ti and thus can be discounted.

In the catalogue of trignometric stellar parallaxes compiled by Jenkins (1952), there is not a single star listed closer than 50 pc down to mag. +9, within 5 deg. of the centre of our preferred search area. This catalogue seems fairly complete down to mag. +9, and contains many stars fainter than this. A star as close as (or closer than) 10 pc would be expected to have a high proper motion (of the order of 1" annually), so unless extremely faint, the likelihood of it being missed seems very small.

The evidence is thus in favour of the star being a supernova. With an apparent magnitude close to -9.5, a maximum distance of roughly 1 kpc is indicated. (At such a high galactic latitude interstellar absorption would be negligible).

Our final preliminary question concerns the recorded period of visibility of the star. Several texts mention a visibility of roughly 3 months. Alī ibn Ridwān is quite specific when he states that the star disappeared "at once" when the Sun came within 60 deg. of longitude of it. However, by this time, from the latitude of Fustat (30°N) the star would be only above the horizon during the hours of daylight. After such a period both a Type I and a Type II supernova would have faded by some 4 magnitudes so allowing for atmospheric absorption the star would appear no brighter than Venus. The visibility of this planet in daylight is very subjective.

Much the same situation would hold in Iraq and China, so that there is a simple enough explanation for the disappearance of the star. However, it does not seem possible to decide on the type of supernova. Lacking extensive data on Type II light curves (see Chapter 8), the lengthy total duration of the star recorded in China (a) cannot be regarded as positive evidence of Type I behaviour.

At such high galactic latitude ($\sim 15^{o}$), interstellar absorption would be low. Thus a peak apparent magnitude of -9.5 would suggest a distance to the supernova of $\sim$1 kpc. A distance as great as 1.3 kpc as suggested by Minkowski (1966) would only be possible assuming zero interstellar absorption, Type I properties, and peak apparent magnitude $\sim$-8.5. We will assume a distance of 1 kpc with uncertainty of about 0.3 kpc.

A radio search of the region of interest was reported by Milne (1971), and his 635 MHz map is shown as Fig. 7.5. The source PKS 1459-41 (G327.6 + 14.5), diameter 34 arc min, is one of two SNRs within the region. The other is the Lupus Loop (G330.0 + 15.0) a large arc ($\sim 3^{o}$ radius) of very low surface brightness. The radio properties are suggestive of an SNR of extreme age ($\sim 10^{5}$ years old), distant about 0.5 kpc and with diameter $\sim$50 pc. The positions of the two SNRs for epoch AD 1006 are shown in Fig. 7.6. The question arises as to whether the Lupus Loop might be a possible remnant of the AD 1006 event, despite its radio properties. If it were at the estimated distance of about 1 kpc for the supernova, then the mean expansion rate of the remnant would need to be >150,000 km s^{-1} - an absurdly high value! To reduce this to an acceptable mean expansion rate for a 1000 year old supernova remnant would require it to be at a distance of no more than 100 pc. Such a nearby outburst would occur on average once per 200,000 years, with an expected absolute magnitude of about -14 or brighter! An object many times brighter than the full Moon would surely have attracted universal attention, and such extreme brightness is certainly not in accord with the historical records of AD 1006. On these grounds, quite apart from the fact that the remnant lies well outside our preferred search area, the Lupus Loop must be entirely

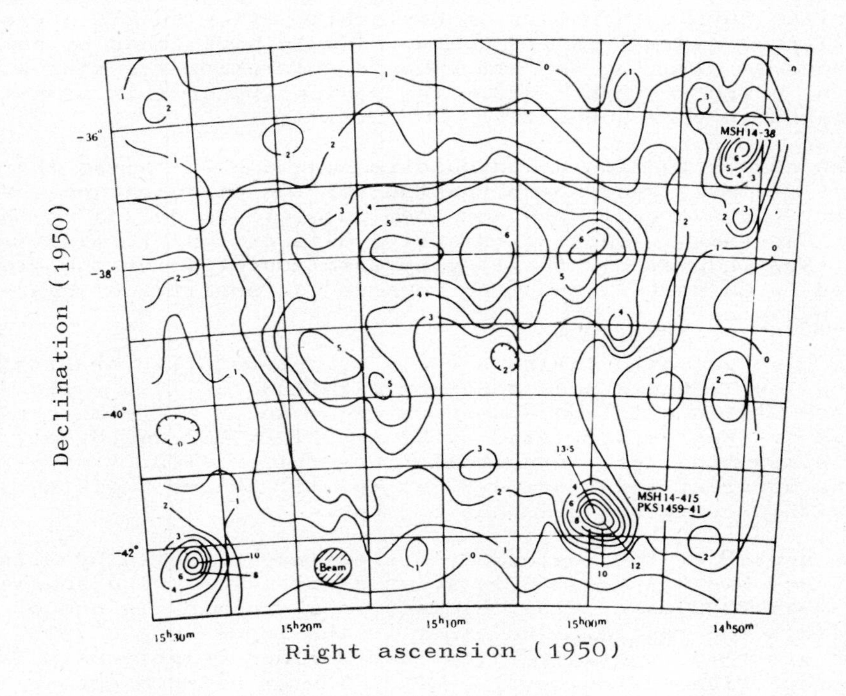

Fig. 7.5. A 635 MHz radio map of the Lupus region (from Milne, 1971).

discounted as a possible remnant of the AD 1006 event.

As noted in Chapter 4 the SNR PKS 1459-41 has the highest galactic latitude of any 'young' remnant, and is only one of four known SNRs lying beyond 10° of the plane. It lies on the periphery of our preferred search area. The chance positional coincidence of a historical new star and a young SNR at such galactic latitude would be extremely remote (less than 1 in 1,000); the available evidence must favour a definite supernova classification for the AD 1006 new star, with PKS 1459-41 as its remnant. This association fits the St. Gallen observations particularly well, and the apparent path of a star at the location of PKS 1459-41 for AD 1006 is shown in Fig. 7.7.

The radio source PKS 1459-41 shows the broken-ring structure characteristic of an SNR (see Fig. 7.8); in addition, a clearly identified non-thermal spectrum ($\alpha = -0.57$) and significant polarization from regions of peak radio brightness confirm a synchrotron origin for the emission. The direction of polarization suggests that the magnetic field is radial, and for a 'young' SNR such a field distribution may occur as a result of

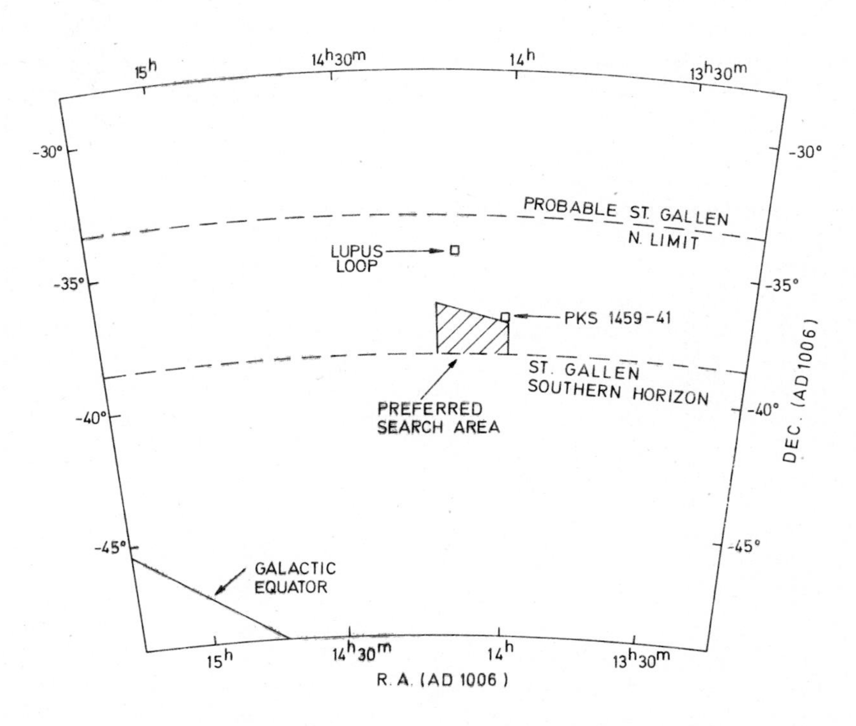

Fig. 7.6 SNRs in the vicinity of SN 1006.

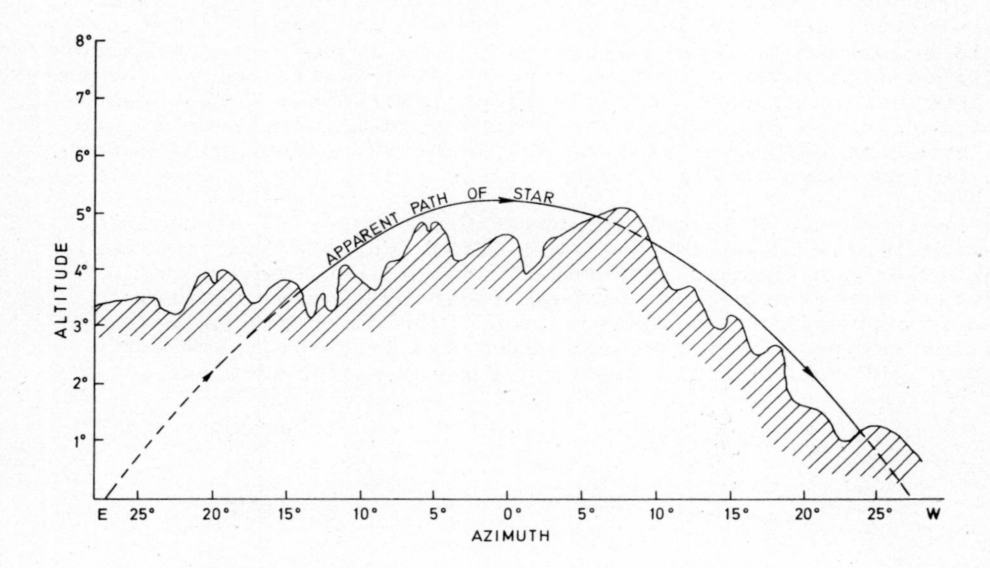

Fig. 7.7. Apparent path of a star at the location of the SNR PKS 1459-41 in relation to the St. Gallen horizon in AD 1006.

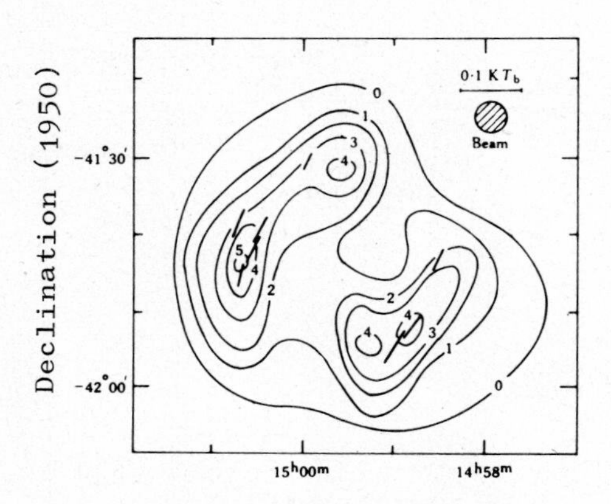

Fig. 7.8. A 5000 MHz map of the remnant of the AD 1006 supernova (from Milne, 1971).

radial stretching of inherent field lines in the expansion of the ejecta. As will be discussed in Chapter 12, the source is of unusually low surface-brightness for its age; however there are other such sub-luminous SNRs, and there are no serious objections to an association of PKS 1459-41 with the AD 1006 supernova.

The optical detection of the remnant has only recently been reported by van den Bergh (1976, IAU Circular No. 2952). Using the 400 cm reflector at Cerro Tololo Interamerican Observatory, delicate wisps of filamentary nebulosity were detected on the north-west periphery of the radio source. Individual filaments are of about 9 arc min in length and from 1 to 8 arc sec in thickness. Van den Bergh suggests that the morphology of the filaments is very reminiscent of that of the optical remnant of the Type I supernova of AD 1572 (see Chapter 10); however as yet it is not possible to confirm this classification on other grounds.

Winkler and Laird (1976) have reported the positive detection of the SNR with the 1 - 10 keV X-ray instrument on the satellite OSO-7; however this detection in X-rays has yet to be confirmed. Nevertheless, on the basis of a thermal origin for the X-ray emission the OSO-7 data suggests a distance of 1.2 kpc to the source, consistent with the distance estimated for the AD 1006 supernova from the historical accounts of its brightness.

It is unfortunate that the remnant of the most spectacular stellar outburst ever recorded by man should have been so poorly studied in modern times. Radio observations have been rather limited, and the optical and X-ray emission only recently detected. Theoretical consideration of the evolution of the source seems almost non-existent. We would completely discount criticism of the classification of the AD 1006 outburst as a supernova, and believe that the association with the SNR PKS 1459-41 must now be taken as certain. It is to be hoped that general acceptance of this association will mean that in future the source will not suffer the observational and theoretical neglect of the past.

Chapter 8

THE BIRTH OF THE CRAB NEBULA

The new star which appeared in AD 1054 was not nearly so bright as its predecessor almost half a century earlier. However, once again we find an impressive list of observations from the Orient.

More than 30 years ago, a detailed survey by Duyvendak (1942) brought to light 5 separate Chinese and Japanese observations of the star. Scarcely any further records have since been discovered, although in recent years the known observations have been subject to much closer scrutiny.

The Crab Nebula, the remnant of the star, has become one of the mos extensively studied objects in astrophysics, but it is perhaps useful to get the original outburst in true perspective. Although well placed for observation in the Northern Hemisphere, there is not a single known record of the star from Europe or the Arab Lands. A Chinese record (see below) implies that the star remained visible for no more than 23 days in daylight, suggesting that it was never much brighter than Venus. It was thus considerably fainter than the new stars of AD 185 and 1006, and was probably of similar brilliance to the supernova of AD 1572 (see chapter 10). Nevertheless, the impact that a number of crude observations made nearly a thousand years ago has had on astrophysical research today seems scarcely believable.

Let us look at the oriental records of the new star. These are all from China and Japan; no astronomical observations of any kind are reported from Korea (in the Koryŏ-sa) in AD 1054 (the 8th year of King Munjong). The reason for this is obscure since the previous year is well covered. Possibly the original observations were lost before the Koryŏ-sa was compiled (there are numerous similar gaps at other periods).

Translations of the Chinese and Japanese records are as follows:

(a) Sung-shih (Astronomical Treatise, chapter 56). "1st year of the Chih-ho reign period, 5th month, (day) chi-ch'ou. (A guest star) appeared approximately several inches (k'o-shu-ts'un) to the south-east of T'ien-kuan. After a year and more it gradually vanished (mo)".

The date of first appearance corresponds to AD 1054 July 4. This report is to be found in the section entitled "Guest stars" in the astronomical treatise, hence the nature of the star is omitted in the text itself. This same account is to be found verbatim in the section on guest stars in the Wên-hsien-t'ung-k'ao, an encyclopedia compiled by Ma Tuan-lin around AD 1280. As this was more than half a century before the Sung-shih was composed, it is apparent that Ma Tuan-lin had access to the same source as the writers of the official history.

(b) Sung-shih (Annals, chapter 12). "(1st year of the Chia-wu reign period, 3rd month) (day) hsin-wei. The Director of the Astronomical Bureau reported that since the 5th month of the 1st year of the Chih-ho reign period a guest star had appeared in the morning at the east, guarding (shou) T'ien-kuan, and now it has vanished (mo)".

This report confirms the month of first sighting given in (a) above and furthermore gives the exact date when the star was no longer visible. This latter date corresponds to AD 1056 April 17, making the total duration more than 21 months. It is interesting to note that this is the only mention of the star in the Annals; there is no reference to it in the discovery year. We might remark that we have here by far the most valuable account of a new star in the Pên-chi (i.e. Imperial Annals) of any dynastic history.

(c) Hsü-tzŭ-chih-t'ung-chien-ch'ang-pien (chapter 176). "(1st year of the Chih-ho reign period, 5th month), (day) i-ch'ou. A guest star appeared approximately several inches to the southeast of T'ien-kuan".

The Hsü-tzŭ-chih ..... is a Sung chronicle written by Li Tao, who died in AD 1184 It thus precedes the Sung-shih by more than 150 years, but even so a similar lapse of time had occurred between the observation of the new star and Li Tao's recording of it. Except that it does not mention the disappearance of the star, the text is virtually identical with that in the Astronomical Treatise of the Sung-shih (a) above. Indeed there are only two slight differences. The Hsü-tzŭ-chih contains an additional character chih, an optional connecting link which does not alter the sense in any way. However, this work gives the cyclical day as i-ch'ou (the 2nd day) of the Sung-shih. Reference to chronological tables (e.g. those of Hsüeh Chung-san and Ou-yang I, 1956) shows that there was no i-ch'ou day in the 5th month of the discovery year, but there was a chi-ch'ou day. However, the written characters chi and i are so similar that they are readily confused. Ho Peng Yoke (1966) cites many such instances occurring in the Chin-shu. Towards the end of the 5th month, the Hsü-tzŭ-chih has successive entries on the 18th, 22nd and 23rd cyclical days, followed immediately by the guest star account under the day i-ch'ou. It seems clear that Li Tao originally had the guest star listed on the day chi-ch'ou (the 26th day), but this was subsequently corrupted to i-ch'ou.

A late summary of the work, the Hsü-tzŭ-chih-t'ung-pien gives the correct cyclical day chi-ch'ou, but without comment. So much for this textual question.

(d) Sung-hui-yao (chapter 52). "1st year of the Chih-ho reign period, 7th month, 22nd day ..... Yang Wei-tê said, 'I humbly observe that a guest star has appeared; above the star in question there is a faint glow, yellow in colour. If one carefully examines the prognostications concerning the emperor, the interpretation is as follows: The fact that the guest star does not trespass against Pi and its brightness is full means that there is a person of great worth. I beg that this be handed over to the Bureau of Historiography'. All the officials presented their congratulations and the Emperor ordered that it be sent to the Bureau of Historiography.

During the 3rd month of the 1st year of the Chia-yu reign period the Director of the Astronomical Bureau said, 'The guest star has vanished (mo), which is an omen of the departure of the guest'. Earlier, during the 5th month in the 1st year of the Chih-ho reign period (the guest star) appeared in the morning in the east guarding (shou) T'ien-kuan. It was visible in the daytime, like Venus. It had pointed rays on all sides (lit. 'in the four directions') and its colour was reddish-white. Altogether it was visible for 23 days".

The Sung-hui-yao ("Essentials of Sung History") was compiled by Chang Tê-hsiang at some time during the Sung Dynasty. The account of the guest star is clearly in two parts, one of which is mainly astrological and the other which gives valuable astronomical details. The first section is very difficult to translate. It seems possible that Yang Wei-tê was using the appearance of the guest star to further his own political motives. This is particularly apparent in the oblique reference to Pi, an asterism more than 15 degrees from T'ien-kuan (see below). His allusion to the colour of the star may be unreliable since as mentioned in Chapter 7, yellow was the imperial colour of the Sung Dynasty. Yang Wei-tê made his speech nearly two months after the appearance of the new star; the date corresponds to AD 1054 August 27.

In the second part of the account, the months of first appearance and final disappearance of the star recorded in the Sung-shih - (a) and (b) above - are confirmed. As the total period of visibility was nearly two years, and the star was not long past conjunction with the Sun when it was first seen (see below), the duration of 23 days must refer to the visibility of the star in daylight. Sightings of Venus in daylight are often recorded in Far Eastern history and it is fairly usual to state the exact number of days that the planet was thus visible.

(e) K'i-tan-kuo-chih (chapter 8). "(23rd year of the Chung-hsi reign period), 8th month, the king died ...... Previously there had been an eclipse of the Sun at midday, and a guest star appeared at Mao. The Deputy Officer in the Bureau of Historiography, Liu I-shou, said, 'This is an omen that (King) Hsin-tsung will die'. The prediction indeed came true".

The K'i-tan-kuo-chih ("Memoirs of the Liao Kingdom"), was written by Yeh Lung-li around AD 1350. Liao was a semi-nomadic kingdom, existing in the extreme north of China from AD 937 to 1125. Duyvendak (1942) pointed out that the king died in the year after the stated year, on a date corresponding to AD 1055 August 28, but he dated the solar eclipse as AD 1054 May 10. This was total in Central China (cf. Oppolzer, 1887) and would be fairly large in the north (there was no other large eclipse until AD 1061). We can obviously no more than guess at the date of the guest star, but judging from the eclipse date it would seem to fit in reasonably well with that in the Sung records. However, Mao is some 20 degrees from T'ien-kuan (for further discussion see below).

Ho Peng Yoke et al (1972) point out that the astronomical details of the above record are copied verbatim in a late work, the Sung-shih-hsing-pien, written about AD 1550, but presumably the compiler had access to the Liao account. N.B. as the work by Ho Peng Yoke et al just cited will be alluded to several times in this chapter, we shall subsequently abbreviate it to HPP .

(f) Meigetsuki (volume 12). "2nd year of the Tenki reign period, 4th month, after the middle decade. At the hour ch'ou (1 - 3 a.m.) a guest star appeared in the degrees of Tsui and Shen. It was seen in the east and flared up (po) at T'ien-kuan. It was as large as Jupiter".

We have already mentioned the Meigetsuki, the diary of the 13th century AD Japanese courtier Fujiwara Sadaie, in chapter 7. The account of the star is to be found under the entry for the 8th day of the 11th month of the Kwanki reign period (AD 1230 December 13). The date of first sighting of the star given in the text corresponds to the days immediately following AD 1054 May 29, but this clashes with the Chinese date. This point will be considered in detail below.

In the above translation we have rendered po as "flared up". It is evident that a verb is intended here. HPP supposed that po implied the compound po-hsing (i.e. a tail-less comet, see Chapter 3), and thus they inferred two separate stars. However, this is highly improbable - we would have expected the full compound, and in any case use of the verb is quite common. Additionally, T'ien-kuan lies "In the degrees of Tsui and Shen" - see below - so that it is clear that the text refers to one and the same star. This was realised by Duyvendak (1942).

(g) Ichidai yoki (volume 1). "2nd year, chia-wu (of the Tenki reign period), 4th month. A great star appeared in the degrees of Tsui and Shen. It was seen in the east and flared up at T'ien-kuan. It was as large as Jupiter".

The Ichidai yoki is a Japanese chronicle of unknown date and authorship. In the original account there are two examples of textual corruption involving very similar characters. These errors were corrected by Kanda (1935). The characters tu ("degrees") and kuan (in T'ien-kuan) are replaced by characters which closely resemble them, but the mistakes are obvious. Otherwise, the Ichidai yoki record is identical with that in the Meigetsuki - (f) above. As the Meigetsuki gives more information (regarding date and time), and is free from error, it is in all probability the earlier source. In any event, the two records are by no means independent, as for (a) and (c) above.

(h) *Dainihonshi* (chapter 359). "2nd year of the Tenki reign period, 4th month, A guest star was seen". This very brief account in such a late Japanese history may well be derived from either (f) or (g).

Summarising the historical sources, we have 5 independent accounts, (a), (b), (d), (e) and (f). As already pointed out, the Chinese sources (a) and (c) share a common origin, as is true of the Japanese sources (f) and (g). In each case the former gives more information so that we shall discard the latter in subsequent discussion. Our object is now to collate the various historical evidence concerning the new star.

Until recently, it was widely accepted that the new star of AD 1054 was a supernova and that the Crab Nebula was its remnant. However, HPP criticised both conclusions. Accordingly, we have felt it desirable to subject the historical material relating to the star to particularly detailed analysis. Leaving a discussion of the exact nature of the star for the present, let us begin by considering the asterisms mentioned. Of the independent sources selected above, (a), (b), (d) and (f) all mention *T'ien-kuan*. It is thus important to establish its identity. Mayall and Oort (1942) in their discussion of the Crab Nebula, believed that it was not possible to locate the asterism "much more precisely than 'near ζ Tauri'". However, HPP made a detailed investigation of early Chinese historical sources and proved conclusively that *T'ien-kuan* is ζ Tau itself, i.e. an isolated star. Judging from the name ("Celestial Gate") two stars would be expected. Of the many works which HPP cite, we have selected two which are most specific, the familiar *Chin-shu* and the *Ling-t'ai-pi-yüan* ("The Secret Garden of the Observatory"). This latter work, originally written by Li Chi-tsai about AD 580, was revised around AD 1050 by Wang An-li, a member of the Sung Astronomical Bureau to incorporate observations made in his own time.

According to the *Chin-shu*, "*T'ien-kuan* is a single star south of *Wu-chê*. It is also called *T'ien-mên* (which also means "Celestial Gate"). It lies where the Sun and Moon move".

The *Ling-t'ai-pi-yüan* is more specific. We read, "*T'ien-kuan* is a single star north of *Tsui*. *T'ien-kuan* is distant 71 'degrees' (*tu*) from the Pole and is within the 1st degree of *Tsui*". HPP point out that Ma Tuan-lin in his encyclopedia, the *Wên-hsien-t'ung-k'ao*, quotes a contemporary Sung source, no longer extant, which gives the polar distance as $71\frac{1}{2}$ degrees - otherwise the description of position is identical with that in the *Ling-t'ai-pi-yüan*. Somewhere a copyist's error has occurred. We should mention here that 1 *tu* is equivalent to 0.9856 degrees; there are 365.25 *tu* to a full circle.

The most useful detail of the *Chin-shu* description is that *T'ien-kuan* lies on (or close to) the ecliptic. *Tsui*, mentioned in the *Ling-t'ai-pi-yüan*, is one of the lunar mansions and its determinant star is λ Ori (see Chapter 2). HPP point out that between AD 1049 and 1053 the co-ordinates of many stars were determined or checked by the Sung astronomers using an armillary sphere; this is possibly the period when the position of T'ien-kuan, as recorded in the Ling-t'ai-pi-yüan, was measured. It would then appear that around the epoch AD 1050 the RA of T'ien-kuan was perhaps $0.5^\circ \pm 0.5^\circ$ in excess of that of λ Ori, and the declination either $+21.0^\circ \pm 0.5^\circ$ or $+20.5^\circ \pm 0.5^\circ$. Applying precession, HPP have reduced the co-ordinates to the epoch AD 1950. The RA is then about $5^h\ 38^m$ and the dec. $+21.7^\circ$ or 21.2°, with an uncertainty of about 0.5 degrees in each co-ordinate. This position is in excellent accord with that of ζ Tau itself (viz. $5^h 34.7^m$, $+21.1^\circ$). As this star is fairly bright (magnitude +3.0), and there are no other stars significantly brighter than magnitude +5 within as much as 8 degrees of it, we can regard the evidence for the identity of ζ Tau with T'ien-kuan as conclusive.

The various oriental star maps discussed in Chapter 2 in every case show T'ien-kuan as a single star, and from the more accurate maps, especially the Jesuit and Su Sung charts, it is clear that the location is the same as that of ζ Tau. Thus we have additional independent evidence (if any were needed).

As we have seen, the Chinese records (a), (b) and (c) mention T'ien-kuan only - in (d) there is an oblique reference to Pi (one of the constituents of which is α Tau). On the other hand, the Japanese record (f) additionally states that the star appeared in "the degrees of Tsui and Shen". These are two adjacent lunar mansions - the 20th and 21st (see Chapter 2). Fig. 8.1 shows the position of T'ien-kuan in relation to Tsui and Shen, and the neighbouring lunar mansion Pi. The boundaries of the zones of RA defined by Tsui are marked. These correspond to the positions of the determinant stars of Tsui and Shen. The position of the Crab Nebula is also shown. To illustrate the relative isolation of ζ Tau, we have shown all stars brighter than magnitude +4.5. From the diagram, it is clear that the Japanese description is quite accurate within its own limitations.

The remaining account (e) from the Liao kingdom mentions only the asterism Mao, i.e. the Pleiades. This is more than 20 degrees from ζ Tau, and it is therefore questionable whether the same star is referred to. However, the Chinese and Japanese records do not mention any other new star near this time. Further, the date of the solar eclipse mentioned in (e) ties in well with the date of appearance of the star near T'ien-kuan. As the record is mainly concerned with the death of King Hsin-tsung, rather than astronomical technicalities, the writer may have given simplified information. The Pleiades are one of the few star groups with which everyone is familiar. If we had had only a single reference to T'ien-kuan, then the discord with the Liao account might have been serious. However, we have ample independent evidence that the new star appeared near T'ien-kuan. We shall subsequently confine ourselves to the three Chinese records (a), (b) and (d), and the Japanese record (f).

Having considered the asterisms mentioned in the various texts and shown that there is general agreement on a single position close to ζ Tau, we may now discuss the nature of the star. Let us begin by considering the exact period of visibility.

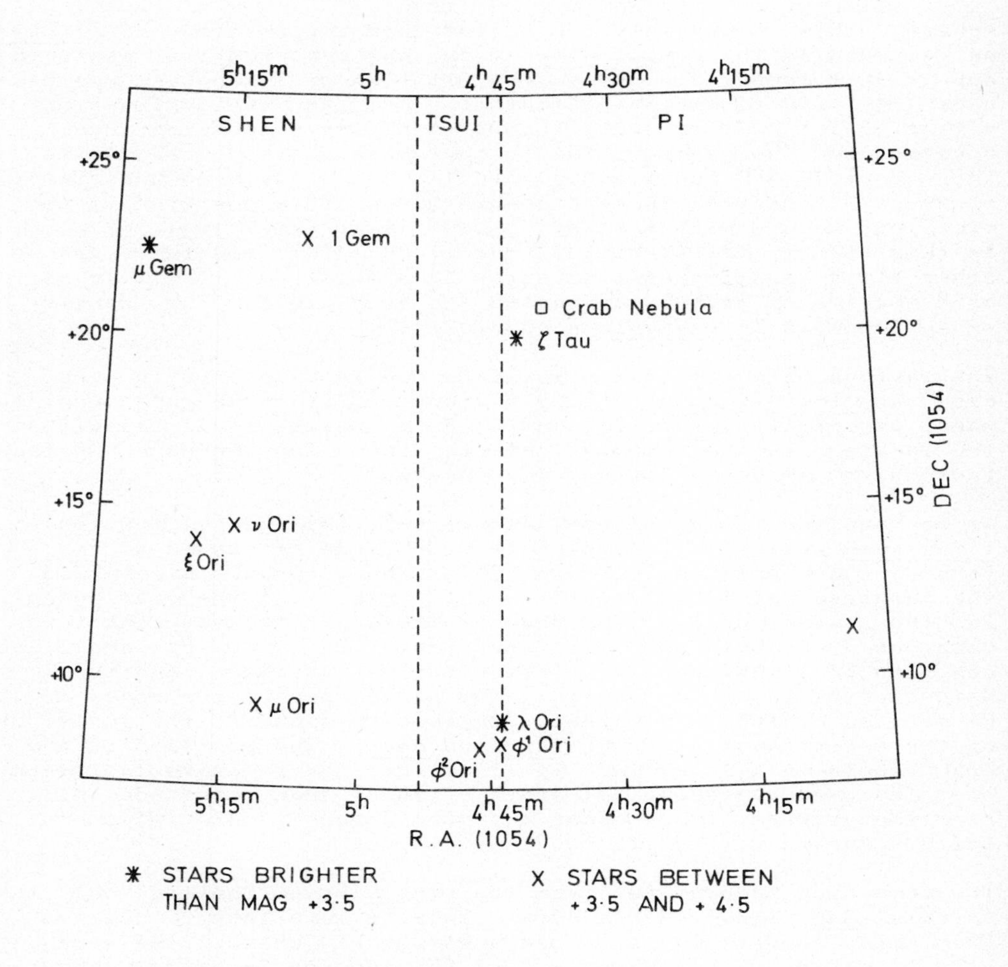

Fig. 8.1. Stars in the vicinity of ζ Tau.

Taking the Chinese records first, (a) gives an exact date of first sighting - the day chi-ch'ou of the 5th month of the Chih-ho reign period, corresponding to AD 1054 July 4. Accounts (b) and (d) confirm the lunar month as the 5th month. On the other hand (a) states only that the star was visible "for a year and more" (which could mean anything up to two years), while (b) gives an exact date of last visibility - presumably the day before hsin-wei of the 3rd month of the Chia-wu reign period, i.e. AD 1056 April 16. We have found in earlier chapters that when Chinese dates are given to the day they are generally correct, and there seems to be no reason for questioning the first and last sighting of the star. For an object which is visible for more than a year, conjunction with the Sun must occur, but we find no mention of heliacal setting or rising. However, there are no means of knowing how much information relating to the star was discarded by the official historians when the *Sung-shih* was written up.

When we turn to the Japanese record (f), we find only a mention of the first sighting, but the date given is much earlier than that in China. The last day of the middle 10-day period in the 4th month of the Tenki reign period corresponds to AD 1054 May 29. According to the text, the star became visible in the days immediately following. Yet it seems inconceivable that the star could have been sighted in Japan a full month before its discovery in China. Additionally, as was pointed out by Duyvendak (1942), ζ Tau would be in conjunction with the Sun around the time of the recorded Japanese sighting (the precise date was May 27). Duyvendak preferred to assume an error of a month in the Japanese record, reading "5th" for "4th", in which case the date of first sighting in Japan would be June 30 onwards. This seems the most reasonable explanation (we have frequently noted errors of a month in earlier chapters). However, as the Chinese dates of discovery and disappearance are precisely given we shall adopt these as definitive.

As far as the nature of the star is concerned, it is significant that both the Chinese and Japanese records avoid the use of the normal terms for a comet. The object was described as a guest star (*k'o-hsing*) in China and Japan, and this could refer to either a nova (or supernova) or a comet without a perceptible tail - see Chapter 3. Further, the various accounts remain silent on any motion of the star during the very long period of visibility which would be unusual for a comet (nearly two years). That the star was still in the vicinity of ζ Tau nearly two months after its first appearance is shown by the account (d) in which the Astronomer Royal discussed the astrological implications of the observation that the star did not "invade" the bright asterism *Pi*. This star group is less than 15 degrees from ζ Tau.

Both (b) and (d) describe the guest star as "guarding" (*shou*) *T'ien-kuan*. Ho Peng Yoke (1966) lists several early definitions of this term, e.g. "to attach and stay by the side of (a constellation or celestial body)", or "to stay (in a particular place) without leaving". Both of these definitions indicate a fixed position. On the other hand HPP, who express doubts regarding the nature of the star, give some importance to an alternative interpretation of *shou* - "moving backwards and forwards without going away". Such an apparent motion could result from the revolution of the Earth, which completed nearly two full orbits while the star was visible, but only if the latter was stationary. We may conclude that the guest star either showed no apparent motion or at most a slight parallactic displacement.

If the new star of AD 1054 was indeed a comet, the extended period of visibility, absence of a tail and lack of any significant motion cannot be reconciled with the observation (d) that it was seen for 23 days in daylight. Daylight comets are by no means infrequent (cf. Chambers, 1909), but they are characterised by a lengthy tail (several tens of degrees), rapid apparent motion at brightest (some 10 degrees daily), short duration in daylight (usually only two or three days at most) and short period of visibility to the unaided eye (a few weeks or months). Ho Peng Yoke (1962) gives detailed translations of oriental records of several hundred comets observed during the last 2500 years. A few of these, described as *hui-hsing* ("broom stars") or *po-hsing* ("rayed stars") were seen for 5 or 6 months, but this is the limit - the maximum recorded duration is no more than 190 days (in AD 254).

Various independent lines of evidence thus concur to disprove a cometary interpretation of the star. Accordingly, we must infer that the object was either a nova or supernova. The "pointed rays" described in (d) as emanating from the star may then be explained as the result of distortion within the eye. This phenomenon is commonly observed with Venus, especially near maximum brightness.

Having established the stellar nature of the new star, we must consider its position in relation to ζ Tau in order to obtain accurate co-ordinates. We find its location described in (a) as "approximately several inches (*ts'un*) to the south-east of *T'ien-kuan*" and in (f) as "at *T'ien-kuan*" (where "at" must be understood as there is no character in the text). Obviously the star must have appeared in close proximity to ζ Tau. The use of a linear unit (*ts'un*) requires clarification. For some reason or other, in Far Eastern astronomy angular separations are regularly quoted in linear units - either *ts'un*, *ch'ih* (equal to 10 *ts'un*) or *ch'ang* (equal to 10 *ch'ih*). The first of these is rather smaller than the British inch. As Kiang (1971) so well remarks, "It may be observed that although in *tu* (i.e. degrees of 365.25 to a circle) we have a perfectly good angular unit, it is never used to express the distance between heavenly bodies, or the lengths of comet's tails, meteor trails or zodiacal light. Instead these linear units are invariably used in such cases".

On the assumption that in the record of the guest star designated as (a) above, the Chinese astronomers were referring to the scale of their armillary sphere, HPP attempted to derive the angular equivalent of the *ts'un* in the text. The armillary sphere constructed by the Chinese astronomer Han Hsien-fu around AD 1000 had a circumference of 18 *ch'ih* (180 *ts'un*). Judging from the list of the dimensions of such spheres throughout Chinese history given by Needham (1959), this figure seems to have been fairly typical. Obviously, on such an instrument 1 *ts'un* would correspond to about 2 degrees.

However, this hypothetical result cannot be correct. Kiang (1971) made a valuable study of 7th century AD Chinese measurements of the position of the ecliptic in relation to the 28 lunar mansions. He derived the result 1 *ch'ih* = $1.50^{o} \pm 0.24^{o}$ (s.e). Thus 1 *ts'un* would be equivalent to about 0.15 degrees. Planetary conjunctions which Kiang analysed, covering a span of more than a thousand years - from the Later Han to the Yüan Dynasty - proved less reliable, but were still consistent with this result within a factor of 2 or so. Our own investigations tend to confirm Kiang's conclusion (cf. Stephenson, 1971), and earlier in this book - in Chapter 5 - we have already discussed a triple planetary conjunction from the Later Han Dynasty (AD 182), where it is apparent that 1 ts'un was roughly equal to 0.1 degrees.

As late as the 17th century, linear units still took the place of angular units in the astronomy of the Far East. This is evident from the Wu-pei-chih ("Treatise on Armament Technology") dated AD 1628,which is quoted by Ho Peng Yoke (1966). What is particularly interesting is the discussion of planetary motions in this work. Fast planetary motion (su-hsing) is defined as "movement of 5 ts'un to 1 ch'ih per day", while haste (chi) is "movement of 1 tu (i.e. degree) per day". Slow planetary motion (ch'ih-hsing) is defined as "movement of 1 or 2 ts'un per day". The mixing of linear and angular units here is curious. Comparison of these rates with the mean daily motions of the planets (of the order of 1 degree for Venus and Mars and 0.1 degree for Jupiter and Saturn) suggests that ch'ih and tu were possibly equivalent.

In order to avoid the seemingly artificial exactness of Kiang's result (without in any way wishing to detract from the value of his investigation) we shall adopt the approximate conversions :

1 ch'ih = 1 tu = 1 degree
1 ts'un = 0.1 tu = 0.1 degree.

We learn from Professor N. Sivin of MIT (private communication) that in his experience ts'un is not an astronomical term. Its usage here is simply one of a great many instances in which the Chinese used ch'ih, and its decimal divisions and subdivisions as a way of expressing non-linear quantities to several decimal places. However we have felt it desirable to discuss the whole question in detail.

The term shu ("several") in text (a) normally means 3, 4 or 5 (occasionally more), so that the "approximately several inches (ts'un)" of the record would imply some 0.3 to 0.5 degree, or perhaps rather more.

HPP raised the question whether this last expression refers to the position of the new star in relation to T'ien-kuan or its apparent physical size. They state that the following interpretation of the record (a) is equally plausible: " .... (a guest star) appeared to the south-east of T'ien-kuan (measuring) about several inches ....". We ourselves cannot agree with this amendment. The text reads much more easily with our original interpretation. Additionally, although many Far Eastern accounts of comets give an estimate of the apparent size, the word ch'ang ("length") is regularly inserted immediately before the estimate. We have noted that this is particularly true of the Sung-shih astronomical treatise.

If we were to adopt HPP's suggestion, this would imply that the apparent angular size of the star was comparable with that of the Moon (0.5 degree). Whether it was a nova or supernova, the star would act as a point source of light. As discussed in chapter 7, all stars appear with the unaided eye to have an apparent diameter which is greater the brighter the star. However, even the excessively brilliant supernova of AD 1006 was only estimated to be 2½ to 3 times the size of Venus, which itself was regarded as having an apparent diameter one-tenth of that of the Moon. We must return to our original interpretation - that the star appeared several *ts'un* away from *T'ien-kuan*. Depending on just how we understand *shu* ("several"), the star could have been anything up to a degree away from ζ Tau, but not significantly more than this.

Regarding the apparent magnitude of the star at brightest, we have only the visibility in daylight of 23 days reported in (d) to go on. The comparison with Jupiter in (f) is of little value since, as was pointed out by Mayall and Oort (1942), at the time the planet was an evening star setting soon after the Sun. The two objects would not be visible simultaneously, for the new star was only observable in the early morning before sunrise. It might be mentioned that Venus was also on the opposite side of the Sun from the star. At the end of the 23-day period of visibility in daylight, the new star would be some 60 degrees west of the Sun and thus was well placed. The magnitude of Venus varies from about -4.5 to -3.5 (except for a very brief period around inferior conjunction). Far Eastern sightings of the planet in daylight are fairly frequent, but are still rare enough to be regarded as omens. This suggests a magnitude of -4 or slightly brighter when the new star ceased to be seen by day.

Some elementary statistical considerations indicate that the probability of the star being a nova (rather than a supernova) is very small. If the star was indeed a nova, it must have been very nearby. From the compilation of nova statistics collected by Payne-Gaposchkin (1957), the absolute magnitude of a nova at maximum varies from about -8.5 for a fast nova to -6.0 for a slow nova. In 23 days a fast nova would fade by at least 3 magnitudes so that the apparent magnitude at maximum would be around -7 or brighter. This would correspond to a distance of less than 20 pc from us. The light curve of a slow nova is very irregular, but variations of a magnitude within 23 days are quite typical. Assuming a peak brightness of -5 leads to a distance of some 15 pc. Let us suppose a maximum distance of 20 pc for a nova as bright as the AD 1054.

Only observations of the brightest novae in recent times are likely to be complete. To obtain some idea of the frequency of novae, let us use the fact that five novae reached an apparent magnitude brighter than +2 during the first 50 years of the present century. From Payne-Gaposchkin (1957) the mean distance of these stars from the Sun was 300 pc. Thus, assuming a nova occurring within 300 pc of the Sun on average once every decade, a nova would occur within 20 pc only about once every 30,000 years, At this small distance away, the distribution of stars is isotropic, whereas ζ Tau is close to the galactic equator ($b = -4^{\circ}$). The probability of a nova occurring so close to the Earth and in such a low galactic latitude during the last 2,000 years (the period of detailed historical records) is of the order of 0.01.

Searches for post novae in the area around ζ Tau have proved negative. As discussed in chapter 1, the generally accepted explanation of nova outbursts is mass transfer in a binary system containing a highly evolved red star and a white dwarf of very high surface temperature. Mayall and Oort (1942) made a survey of the area around ζ Tau for post novae and found none. We have decided to approach the question using the large estimated parallax of the new star if it was in fact a nova.

At such a small distance, proper motion is likely to be large, typically of the order of 0.2" per annum. We have checked a detailed star catalogue containing spectra and proper motions for stars down to magnitude +19. Of the few stars within 2 degrees of ζ Tau brighter than magnitude +8, only two have proper motions greater than 0.025". These are a star of spectral type F8 and magnitude +7.5 (proper motion 0.10") and a star of type A2 and magnitude 6.4 (proper motion 0.05"). From its intermediate spectral type, the former is certainly not a post nova, while the latter has a spectroscopic parallax of 0.010" (Schlesinger, 1940), implying a main sequence star.

The catalogue of trignometric parallaxes compiled by Jenkins (1952) seems fairly complete down to magnitude +8, and contains many stars fainter than this. Within 2 degrees of ζ Tau there is only a single star listed closer than 50 pc. This is of spectral type G4 and magnitude +8.6. It has a well determined parallax of 0.053" ± 0.011", indicative of a main sequence star. Such a catalogue is likely to be particularly complete for nearby stars since these reveal themselves by their large proper motions.

Finally, the catalogue of stellar parallaxes (many of them spectroscopic) produced by Schlesinger (1935) covers much the same range of magnitude as Jenkins' work, but contains no extra stars nearer than 50 pc. We might mention here that ζ Tau itself (magnitude +3.0) is of type B3 spectrum, but very small parallax -0.14", determined spectroscopically. The star is thus a main sequence blue star, clearly not a post nova.

We can conclude that if the star of AD 1054 was a nova, its range of brightness was at least 13 magnitudes. This is greater than for all but the very fastest nova. Only two such stars are known to have exceeded this range and both were extremely fast. Nova Puppis (1942) reached a peak of +0.4 and the post nova is now around +17 (cf. Payne-Gaposchkin, 1957). Nova Cygni (1975) reached a magnitude of +1.8 (cf. Isles, 1976). At the present time it is still declining, but a search for the pre-nova down to magnitude +20 has so far proved unsuccessful. In 23 days Nova Puppis declined by 5 magnitudes, and Nova Cygni by as much as 7. The peak magnitude for the AD 1054 star would be -9 or -11 on this basis, comparable with the Moon in brightness. That the star was a very fast nova of this type can be ruled out by the lack of any known sighting in Europe and the Arab Lands, and especially by the absence of any reference to unusual brilliance in the Chinese and Japanese sources. Bearing in the mind the low probability of the event being a nova in any case, we feel the nova hypothesis must be rejected. We have no hesitation in concluding that the new star of AD 1054 was a supernova.

The remnant of the supernova of AD 1054 has long been accepted to be the Crab Nebula (particularly since the work of Mayall and Oort, 1942). The Crab is an established supernova remnant. Let us discuss the evidence in favour of and against this identification.

Our usual estimate of the maximum distance of the supernova can be made assuming a value for peak absolute magnitude. Unlike novae, whose rise to maximum brightness is usually very steep, supernovae have a rounded peak in the light curve. This makes discovery several days before the maximum probable. Once more using the 23-day period of visibility in daylight, a fair estimate of the apparent magnitude at maximum would be -5. Allowing 1 magnitude per kpc for interstellar absorption (Minkowski, 1968, used the more precise figure of 1.07 for absorption in the direction of the Crab Nebula), and a peak absolute magnitude of -19 gives a distance of 2 to 2.5 kpc. As discussed below, the extrapolated date of outburst of the Crab Nebula from the rate of expansion of the optical filaments is AD 1140$\pm$10. This is in fair agreement with the date of the star since acceleration of the expansion has probably taken place (see later). The distance of the Crab Nebula ($\sim$2 kpc) is in excellent accord with our rough calculation for the supernova. Additionally, the angular distance of the nebula from ζ Tau (1.1 degree) fits in well with that for the new star (about a degree at most). Proper motions of the nebula and ζ Tau are negligible.

The one problem arises from the fact that the Crab Nebula lies to the north-west of ζ Tau, whereas according to text (a) the star appeared to the south-east of it, i.e. in the very opposite direction. HPP believed that this was a serious obstacle to the identification but they were of the opinion that sources (a) and (c) were independent. We have shown that this is not the case. As the contents of the Sung-shih have probably suffered much at the hands of the Yüan Bureau of Historiography, we need not be overconcerned at this apparent discrepancy, particularly as three other texts (b), (d) and (f) confirm that the star appeared in the vicinity of ζ Tau. Again, the astronomer who made the original observations may have mistaken the relative positions of the new star and T'ien-kuan in writing up his report. To quote a parallel example, on a date corresponding to AD 1253 September 16, the astronomical treatise of the Koryŏ-sa reports that Mars occulted the south-east star of Yü-kuei. This asterism in Cancer consists of only four stars, each some 5 degrees apart. Our calculations show that in place of the south-east star (δ Can) we must read the north-west star (η Can). Obviously this merely indicates that such confusion can occur.

If we do not accept this interpretation, we have to explain why the supernova which did produce the Crab Nebula is not recorded, and also what became of the remnant of the AD 1054 star. From the expansion rate, the date of appearance of the star which produced the Crab Nebula was certainly within a century of AD 1054. At the known distance of the nebula ($\sim$2 kpc), if the star was a supernova of Type I its peak brightness would be around magnitude -5.5, and if of Type II around -4. The worst possible conditions for visibility of the star would occur if it first appeared a few days or weeks before conjunction with the Sun. Even allowing as much as 50 days between the initial outburst and the eventual discovery after conjunction with the Sun (which seems excessive), fading in the intervening time would be only some 3 magnitudes for a Type I supernova and 1.5 magnitudes for Type II. The star would thus be *at least as* bright as Jupiter when it was first seen.

The sudden appearance of such a prominent star almost exactly on the ecliptic (the most extensively observed region of the sky) could scarcely fail to be noticed. If no planet happened to be near, the monthly conjunction of the Moon with the star undoubtedly would draw attention to it. Yet there is not a single record in oriental history which might be taken to refer to the star except in AD 1054, and for this we find several Chinese *and* Japanese descriptions.

If we are in fact dealing with two separate supernovae, we have to explain the remarkable coincidence of two such stars appearing within less than a century of one another, only about 1 kpc apart, and at a mean distance of 2 kpc from the Earth. Furthermore, these two stars appeared roughly in our line of sight. The probability of such a double event is minute - of the order of 10^{-5}.

Again, what of the remnant of the AD 1054 star? A young supernova remnant at a distance of 2 - 2.5 kpc at most would be expected to be detected as a radio source. However there is no known supernova remnant within 5 degrees of ζ Tau apart from the Crab Nebula itself, and within 20 degrees the only 3 SNRs (PKS 0607+17, S147, IC443) are all of extreme age.

We feel that the various lines of evidence discussed above justify the reversal of the recorded direction of the guest star of AD 1054 with respect to ζ Tau as recorded in text (a), and permit reinstatement of the Crab Nebula as one of the few select remnants of precisely known age.

HPP, by drawing attention to the problem of the relative directions of ζ Tau and the Crab Nebula, and other supposed difficulties which we have answered above, were no doubt expressing sincere reservations. However, their paper is based almost entirely on destructive criticism with very little constructive effort (apart from the valuable identification of *T'ien-kuan*). Such a biased approach is contrary to the interests of science. Ho Peng Yoke's contribution to the field of oriental astronomy has been outstanding, but in this particular case we feel that his efforts have been misguided.

While still on the subject of the historical records, we might mention that Miller (1955a and b) drew attention to two possible rock art records of the AD 1054 supernova. He believed that these represented a close conjunction of the Moon with the star and calculated that such a configuration took place on the morning of 5 July in AD 1054. This is very soon after discovery in China. Miller's suggestion was enthusiastically taken up by Brandt et al (1975), who provided further evidence, although they pointed out that it was not possible to prove their conclusions. Archaeological findings indicate that the period is roughly correc (within a century or so). Our own reaction to this fascinating hypothesis is favourable. However, we emphasise that the pictographs tell us nothing about the star. On the other hand, Ellis (1975) stresses that the Pueblo people "characteristically did not record exciting events" ,and gives several examples. She believes the pictographs to be nothing more than clan symbols. At the moment, the question seems open, but even a satisfactory interpretation does not affect our analysis.

It should be noted that from the historical data it is not possible to decide whether the supernova was of Type I or Type II. All we know about the light curve is that the star reached a magnitude close to -4 around 27 July in AD 1054 and disappeared from sight on or about 16 April in AD 1056. When the star finally faded, it was some 40 degrees east of the Sun, and was thus well placed for visibility, especially in view of its northerly situation. This sets a limit of about +5, or a little fainter some 21 months after it reached -4. Type I light curves have been followed for about this length of time, but not Type II (cf. Zwicky, 1965). No Type II supernova seems to have been followed for more than about 150 days or so, and the form of the light curve after this time is entirely hypothetical. A clear distinction could be made if more data on Type II light curves could be obtained over a longer period, and we regard this as a matter of some urgency.

For some considerable time the AD 1054 supernova was classified as a Type I event. As Minkowski (1971) has pointed out this classification was actually something of an historical accident. At the time of early suggestions of an association between the Crab Nebula and the AD 1054 outburst (for example Mayall and Oort, 1942), similarities with the supernova in the galaxy IC4182 were emphasized. This latter supernova was later to become the prototype of Type I supernovae, and the assignment was automatically (but without any sound basis) tagged to the supernova of AD 1054. There is in fact ample evidence that the AD 1054 supernova could not have been a simple Type I event (Minkowski, 1966; 1968).

An often quoted observation of Geoffrey Burbidge is that modern astronomical investigations have been equally divided between studies of the Crab Nebula and studies of everything else in the Universe. While this is undoubtedly a gross exaggeration, it does highlight the unique place of the Crab Nebula in astronomy. No other object has stimulated so many new theories, or so readily provided the observational means to test them. It was the first galactic object recognised to be the remnant of a supernova; the first radio source and the first X-ray source to be identified with a particular object (other than the Sun); the first source for which linear polarisation, indicative of synchrotron emission, was observed; the first (and to-date only) remnant of an historically observed supernova for which a pulsar has been detected; the first SNR with an optically observed pulsar. The Crab Nebula is one of only two SNRs displaying extended radio, optical, and X-ray emission plus a pulsar, (the other is the Vela SNR);

but despite all of these features, and intense observational and theoretical study, it remains a unique and enigmatic object displaying few of the common evolutionary properties of nearly all other SNRs.

Optically the Crab Nebula must be one of the most familiar of stellar objects. The discovery of the nebulosity has been attributed to an English amateur astronomer John Bevis in 1731. It was 'rediscovered' in 1758 my Messier in his observations of the passage of Halley's comet in that year, and was listed as M1 in his famous catalogue of objects that might cause confusion in cometary studies (the first catalogue of nebulae). The common name of the Crab Nebula seems to have been acquired about a century later, and may have arisen from an 1844 drawing of the object by Lord Rosse which resembled a crab — certainly the name has been in general use from about that time. The first photograph of the Crab Nebula was obtained by Robertson in 1892 using a 20 inch reflector; Plate 9 shows a 1952 photograph using the Palomar 200 inch reflector.

Early spectral observations of the nebula were by Mayall (1937) and others; these indicated that the predominant emission lines were O II , O III , N II , S II , and the Balmer series of hydrogen. A distinctive feature was the strength of the continuum emission. Subsequently Minkowski (1942) detected relatively weak He lines. He isolated all the emission lines to a system of fine filaments, and in fact the nebula is now recognised to have two dissimilar components; an amorphous, bluish region, plus multi-coloured filaments which are predominantly green near the centre of the nebula. The blue region is generated directly by the synchrotron process - it is believed that the electrons are accelerated to the speeds required to emit synchrotron radiation in the strong magnetic field of the active pulsar. Polarization of this component, indicative of its synchrotron origin, was first demonstrated by Dombrovsky (1954).

The Crab Nebula remains the only SNR with optical emission unambiguously identified as synchrotron in nature. (The optical emission from all other SNRs is thought to result from shock-heated interstellar material.) The synchrotron emission extends into the far ultra-violet, and this radiation ionizes the gas filaments which are then seen in the emission lines characteristic of their chemical composition. The more energetic synchrotron radiation photons are believed responsible for doubly ionizing oxygen producing the characteristic green lines seen near the centre of the nebula. The exact interpretation of spectral data in terms of chemical composition remains open to some debate. An unexpected result of Davison and Tucker (1970) is that the abundance ratio of helium to hydrogen is a factor of from 6 to 10 larger than normal, providing evidence of nuclear transformations in supernovae.

Comparison of line emission photographs of the Crab Nebula taken ten or more years apart show that the object is expanding. As noted earlier, the time scale of this expansion is compatible with its origin in the AD 1054 event, with its centre within about 10"

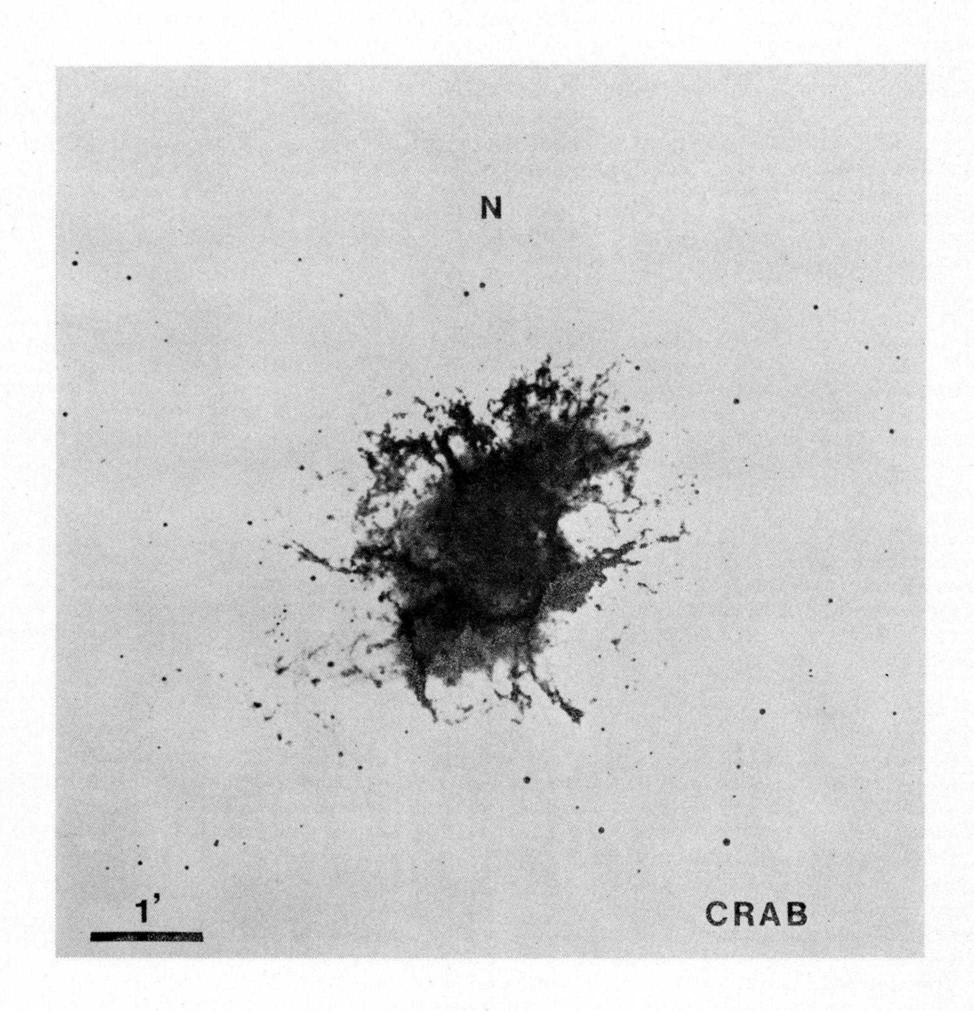

Plate 9. The Crab Nebula.

of the pulsar.

The filamentary expansion was discovered by Duncan (1921), with further investigation by Duncan (1939) and Baade (1942). More recent observations have not greatly affected the interpretation of these early results. The angular expansion in the direction of the major axis of the nebula was determined to be 0.235"±0.008" per year, which for the size of the nebula and assuming a uniform expansion rate places the date of the outburst as AD 1140±10 . In view of the certain association of the Crab Nebula with the AD 1054 supernova, this anomaly must be interpreted as indicating an acceleration over the history of the nebula - if constant, this acceleration amounts to 0.0014 cm/sec^2 for the assumed distance of the nebula, with the initial expansion velocity being $\sim$1700 km/sec along the major axis of the nebula and $\sim$1100 km/sec along the minor axis. Trimble (1971) has suggested that this outward acceleration could be imparted by the outward pressure of a magnetic field and/or a relativistic particle flux with values comparable with minimum estimates required to produce the observed synchrotron radiation.

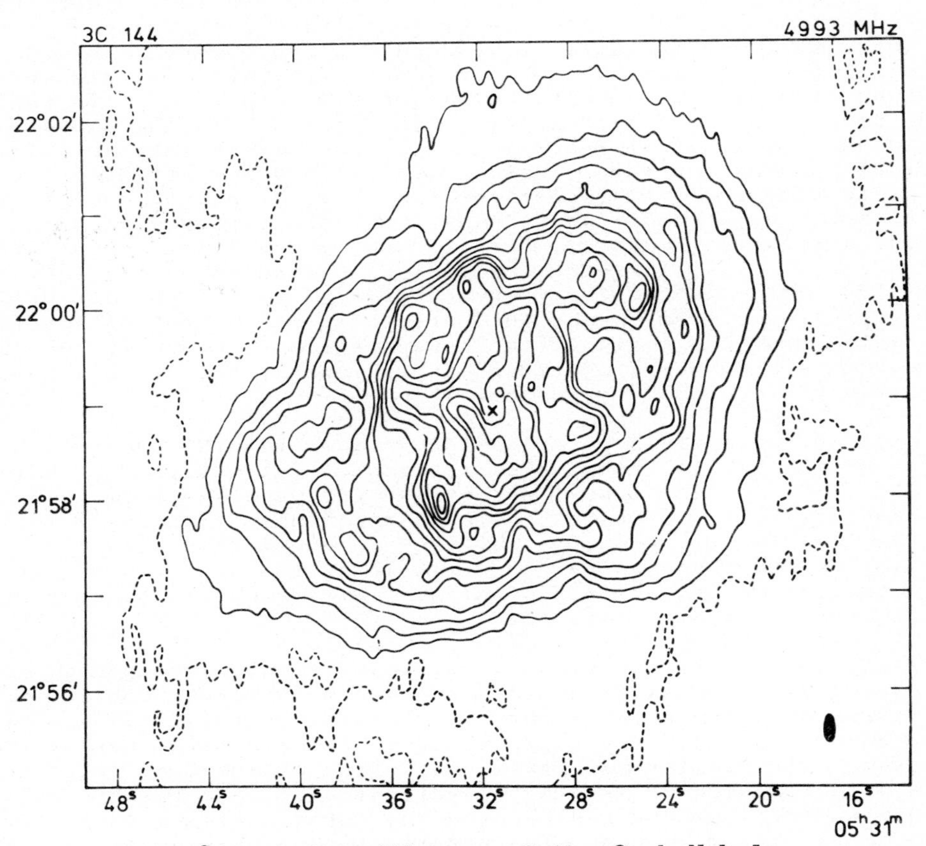

Fig. 8.2. A 5000 MHz map of the Crab Nebula, from Wilson, 1971. The cross marks the pulsar.

The association of the Crab Nebula with the strong radio source Taurus A was confirmed by the observations of Bolton and Stanley (1949), this being the first positive identification of a discrete cosmic radio source with an optical object. Since that time the Crab Nebula has probably been subjected to more intense observation than any other radio source, so that today its spectral and structural properties are well understood. High-resolution maps (see Fig. 8.2) show that the radio distribution is remarkably similar to that of the optical continuum, although larger by a factor of $1\frac{1}{2}$ to 2, depending on direction. The distribution is approximately centred on the Crab Pulsar NP 0532, first detected at radio wavelengths by Staelin and Reifenstein (1968). It is essentially amorphous with no hint of the peripheral brightening characteristic of the vast majority of radio remnants of supernovae. Many of the bright ridges of radio emission appear to be associated with the bright optical filaments; Wilson (1971) has suggested that this could be explained by a magnetic field enhancement of 1.5 or more near the filaments.

The radio spectrum of the Crab Nebula has been thoroughly investigated. Although observations extend from 10 MHz to 250 GHz (see Fig. 8.3) the most accurate measurements are in the range 0.1 to 10 GHz. At lower frequencies ionospheric and inter-planetary scintillation effects make observations difficult; at higher frequencies calibration difficulties could help to explain the wide dispersion in flux density estimates. Within the range 0.1 to 10 GHz the spectral index $\alpha = -0.26$, somewhat flatter than typically found for SNRs. Baldwin (1971) has investigated the complete electromagnetic spectrum (see Fig. 8.4) and shown that it fits a simple smooth curve over the whole observed range, having a spectral index of -0.26 for radio wavelengths (10^7 to 10^{11} Hz), -0.9 for infrared and optical wavelengths (6×10^{13} to 10^{15} Hz), and -1.2 for X-ray wavelengths (2×10^{17} to 10^{20} Hz). The overall spectrum may be interpreted as a synchrotron spectrum extending from the radio to the X-ray region (Shklovsky, 1966), requiring a permanent injection of high-energy electrons with total energy of $\sim 10^{38}$ ergs/sec.

In contrast to optical polarisation greater than 50 percent in parts of the nebula, radio polarisation (first detected by Mayer et al, 1957) is less than 1 percent at frequencies below 1 GHz, increasing slowly to about 7 percent at 10 GHz; Faraday rotation in the shell of filaments is believed to obliterate the polarisation at low frequencies. Nevertheless, the observed degree of polarisation is entirely consistent with a wholly synchrotron origin for the radio emission.

At X-ray wavelengths the Crab consists of a pulsed component (clearly associated with the pulsar) and an extended continuum emitter. Even prior to the advent of X-ray astronomy SNRs were proposed as likely X-ray sources due to emission from heated swept-up interstellar material. Detection of the Crab Nebula as a strong source of X-rays (Bowyer et al, 1964) was therefore hardly unexpected, and represented the first positive identification of a cosmic X-ray source. The nature of the X-ray emission was however to provide many surprises. The extended continuum emitter

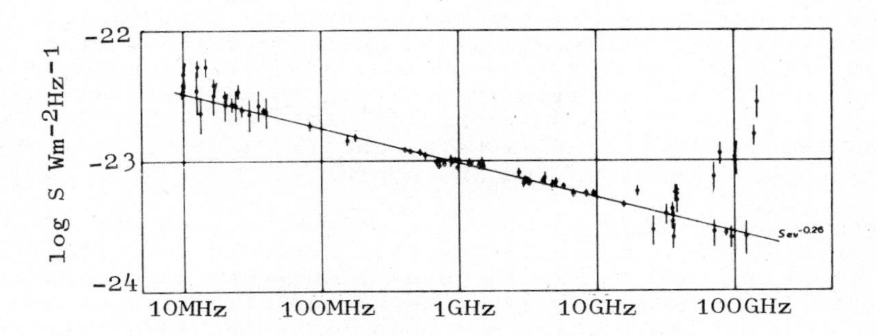

Fig. 8.3. The radio spectrum of the Crab Nebula (from Baldwin, 1971).

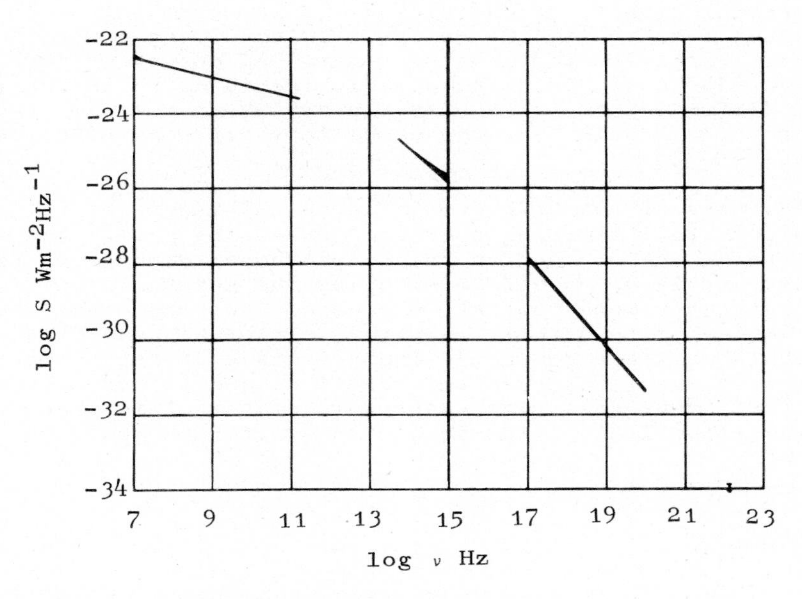

Fig. 8.4. The electromagnetic spectrum of the Crab Nebula (from Baldwin, 1971). The thickness of the line covers the limits of uncertainty at any frequency.

is about 60" by 90" in extent (Hawkins et al, 1974), with a well determined power-law spectrum indicating radiation by the synchrotron process rather than the expected thermal emission from a hot gas. The observation by Kestenbaum et al (1971) that the X-ray emission is polarised confirms that high-energy electrons are almost certainly responsible for the emission. In this regard the Crab is unique, for there are at the present time no other known SNRs for which the synchrotron process dominates the X-ray emission Extended soft X-ray emission from beyond the central extended component has recently been detected; this emission may result from the passage of the shock wave, and the significance of this observation will be discussed in Chapter 12.

The Crab pulsar NP 0532 is the only pulsar to have been found at optical radio, and X-ray wavelengths. Prior to the detection of pulsed emission, the source had been identified as a compact radio object with steep spectrum by Hewish et al (1964) and as a 16th-magnitude star, which had been suggested to be the stellar remnant of the supernova as in the scenario of Zwicky (1939). The pulsar period of 0.0331 second is gradually increasing at a rate of 3.5×10^{-8} second per revolution per day; these values are completely compatible with the pulsar's formation about 1000 years ago with a doubling of the period in that time, and provide yet another link between the Crab Nebula and the AD 1054 new star. The discovery of the pulsar, and the realization that it could provide a continuing supply of high-energy electrons, solved the long-standing problem of explaining the large amount of energy still being radiated by the entire nebula so long after the outburst. The estimated loss of rotational energy of the pulsar of 5×10^{38} ergs/sec is comparable to the total radiative energy losses, confirming that the pulsar provides an adequate energy source for the Nebula over the complete electromagnetic spectrum (Woltjer, 1971).

The uniqueness of the Crab Nebula is no doubt sufficient justification for its intensive investigation in the past. However, despite being one of the few known supernova remnants of precisely known age, it is of no use in studying the general properties and evolutionary behaviour of SNRs, since not one other galactic SNR resembles it in detail. For this reason we feel that the Crab Nebula may have dominated the attention of astronomers for too long; we would express the hope that the rather more typical remnants of other historical supernovae may in future enjoy the observational attention so lavishly afforded the Crab.

Chapter 9

THE INVADER OF THE GUEST HOUSES

The guest star which appeared in AD 1181 probably aroused more interest than any other new star which was seen before modern times, with the exception of the brilliant supernovae of AD 1006, 1054, 1572, and 1604. The star was observed independently in north and south China (the Kin and Sung empires) and also in Japan. As well as being mentioned in several Japanese histories the guest star was reported in the diaries of certain courtiers living at the time.

We have consulted the original Japanese records in the references cited by Kanda Shigeru (1935), as well as the original Chinese records. Translations of these are as follows:

(a) Sung-shih ("History of the Sung Dynasty" ,chapter 56).
"On the day chi-szŭ in the sixth month of the eighth year of the Ch'un-hsi reign period (AD 1181 August 6) a guest star (k'o hsing) appeared in K'uei-hsiu and invading (fan) Ch'uan-shê until the day kuei-yu of the first month of the following year (AD 1182 February 6), altogether 185 days; only then was it extinguished (mieh)."

(b) Kin-shih ("History of the Kin Dynasty",chapter 20).
"On the day chia-wu in the sixth month of the 21st year of the Ta-ting reign period (AD 1181 August 11) a guest star was seen at (yu) Hua-kai altogether for 156 days; then it was extinguished".

(c) Meigetsuki (Diary of Fujiwara Sadaie, AD 1180-1235)
"On the day kêng-wu, the 25th day of the sixth month of the first year of the Yōwa reign period (AD 1181 August 7) a guest star appeared at the north near (chin) Wang-liang and guarding (shou) Ch'uan-shê."

(d) Dainihonshi ("History of Great Japan"; written AD 1715). The account is identical with that in the Meigetsuki, and is presumably derived directly from this earlier work.

In the following Japanese description we have omitted the date when it is identical with that in the Meigetsuki record. We are grateful to Dr. M.I. Scott, of the University Library, Cambridge, for supplying the notes shown in parentheses.

(e) Gyokuyō (Diary of Fujiwara Kanezane, AD 1164-1200, courtier of the imperial court).
"On the 28th day (commentary:kuei-yu)it was heard that since the 25th day a guest star had been present in the inner sky (commentary: beside Wang-liang), a sign of abnormality indicating that

at any moment we can expect control of the administration to be lost."

(f) Hyakurenshō (History of the imperial court from AD 967 to 1259. Written about AD 1260).
"A guest star was seen at the north pole."

(g) Azuma Kagami (History of the Kamakura military government, AD 1180-1266. In two parts: mid 13th and mid 14th century.)
"At the hour hsü (19^h to 21^h local time) a guest star was seen in the north-east. It was like Saturn and its colour was bluish-red and it had rays. There had been no other example since that appearing in the 3rd year of Kankō (AD 1006)".

The record appears to be incomplete. The NASA translation of the paper by Hsi Tsê-tsung and Po Shu-jen (1965) prefers "it was as large as Saturn", while Shinoda Minoru (1960) renders the passage "it was the colour of Saturn". The translation of Yang (1966) which reads "it was as big as Tsung Hsing (110,111 Herculis)", is quite erroneous.

(h) Kikki (Diary of Yoshida Tsunefusa, AD 1140-1200, a courtier of the imperial court).
"The twenty-fifth day kêng-shen of the eighth month (AD 1181 September 26). Today was the occasion for making auspicious offerings for a good harvest. A guest star had been seen in the NNE and at the present time it still has not faded away."

The star does not seem to have attracted attention in Korea. The Koryŏ-sa ("History of the Kingdom of Koryŏ") contains detailed astronomical records for this period (12th year of Myŏngjong) but the guest star is overlooked.

The oriental asterisms mentioned in the accounts (a),(b),(c) and (d) are all in the vicinity of Cassiopeia. This location is confirmed by the Azuma Kagami record (g) which states that the guest star appeared on August 7 between 19^h and 21^h (local time) in the north-east. As the Sun was then at R.A 9^h 30^m, the star (sighted in China the night before) would probably be observed soon after 19^h 30^m local time, when darkness fell. By this time Cassiopeia (R.A about 0^h) would be considerably to the east of the pole. The observation quoted in the Hyakurenshō (f)-"at the north pole"- gives only a general indication of position.

In the Azuma Kagami account (g) the guest star was stated to be the only example since the brilliant supernova of AD 1006. To what extent comparison should be made is difficult to decide owing to the interval of nearly two centuries between the observations. The "rays" described are presumably merely the protuberances which appear surrounding all bright stars owing to distortion of light within the eye. That they should be so noticeable suggests that the guest star was considerably brighter than zero magnitude. the silence of all the other accounts on the question of the brightness of the star and the absence of reference to it from Korea do not seem to support this conclusion. However it is difficult otherwise to account for the considerable interest aroused by the appearance of the star in both China and Japan.

The comparison with Saturn in the Azuma Kagami record is rather unexpected. At the time of discovery in Japan (19^h to 21^h local time) the only planet visible was Mars, fairly high in the south-east. Jupiter would become visible near midnight, while Saturn, 17 degrees West of the Sun , would only be seen for a short time before sunrise. The magnitude of the planets were about -1,-2, and +1 respectively. Perhaps the guest star (which was visible throughout the night) most nearly resembled Saturn in brightness. There is no possibility of the characters representing the various planets being confused. It is most improbable that the reference is to the similarity in colour between the guest star and the planet even though the low altitude of Saturn may have produced colours of atmospheric origin. The description "bluish-red" needs no qualification, and the observers would have been well aware that the planet is normally whitish in colour.

The guest star was discovered in Japan one day after the first observation in southern China. Possibly the delay of five days in northern China was the result of unfavourable weather conditions. Kepler's supernova was discovered about eight days before the maximum, during which time the brightness increased by some 3 magnitudes (Baade,1943). Assuming that the star of AD 1181 was about as bright as Saturn at discovery, it is possible that the star became significantly brighter than magnitude 0 a few days after discovery.

The discrepancy between the periods of observation recorded by the Sung and Kin astronomers may be significant. The guest star remained visible in southern China for 24 days longer than in the north. As this was the dry season in the north, the inference is that the brightness remained near the unaided eye limit between the last Kin observation and the last Sung observation. This feature is characteristic of a Type I supernova.

On the assumption that the AD 1181 guest star was a supernova of Type I, we can attempt a deduction of the magnitude at maximum. At the time of disappearance the star was approaching conjunction with the Sun, but the high declination (about +60 degrees in AD 1181) would permit two hours of observations at an altitude of more than 30 degrees each morning before nautical twilight commenced. The star is unlikely to have been observed after the magnitude fell below +5.5; it would then be just another faint star on the verge of visibility. Tycho Brahe, under more favourable circumstances, ceased observation of the AD 1572 supernova in Cassiopeia when it became fainter than about +5.3 (Baade,1945). The Type I light curves indicate a fall of between 5 and 5.5 magnitudes in about 180 days. This suggests an apparent magnitude close to zero at maximum, in fair agreement with the other evidence. The magnitude corrected for interstellar absorption may have been considerably brighter than this.

The oriental asterisms Wang-liang, Hua-kai, and Ch'uan-shê, together with Ko-tao which is not mentioned in the accounts of the guest star, are adjacent to one another in Cassiopeia and cover an area approximately 15 degrees square. Apart from the new star of AD 1181, the only other long-duration stars from Table 3.2 which were seen near these asterisms were those of AD 1572 and AD 1592. The new star of AD 1572, which appeared beside T'sê-hsing (propably κ Cas) was Tycho's supernova. The two guest stars of AD 1592 (labelled (B) and (C) in Table 3.2) were considered earlier in Chapters 3 and 4, where it was concluded that they were probably slow novae.

Fig. 9.1 is a detailed chart of the Cassiopeia region for the epoch AD 1900.0. The positions of all stars brighter than visual magnitude +6.0 in the zone R.A 22^h30^m to 3^h30^m, dec. +55° to 75° are shown. Numbers represent magnitudes - for the few variable stars the range is given. Bayer stars are denoted by the appropriate Greek letter. Only a few stars have annual proper motions in excess of about 0.1". These are β Cas, η Cas, δ Cas, and two stars of magnitude +5.7; BD 56° 2966 (R.A 23^h8^m, dec. +57°) and BD +63° 238 (1^h40^m,+63°). The proper motions are in order: 0.5", 1.2",0.3",2.1", and 0.6". The aspect of this part of the sky has thus remained virtually unchanged since the appearance of the guest star. The lunar mansion boundaries at AD 1181, corrected for precession to 1900.0, are indicated by broken lines on Fig. 9.1. The epoch 1900.0 was chosen merely for convenience since the coordinates of the numerous stars represented on the chart were taken from a catalogue for this epoch.

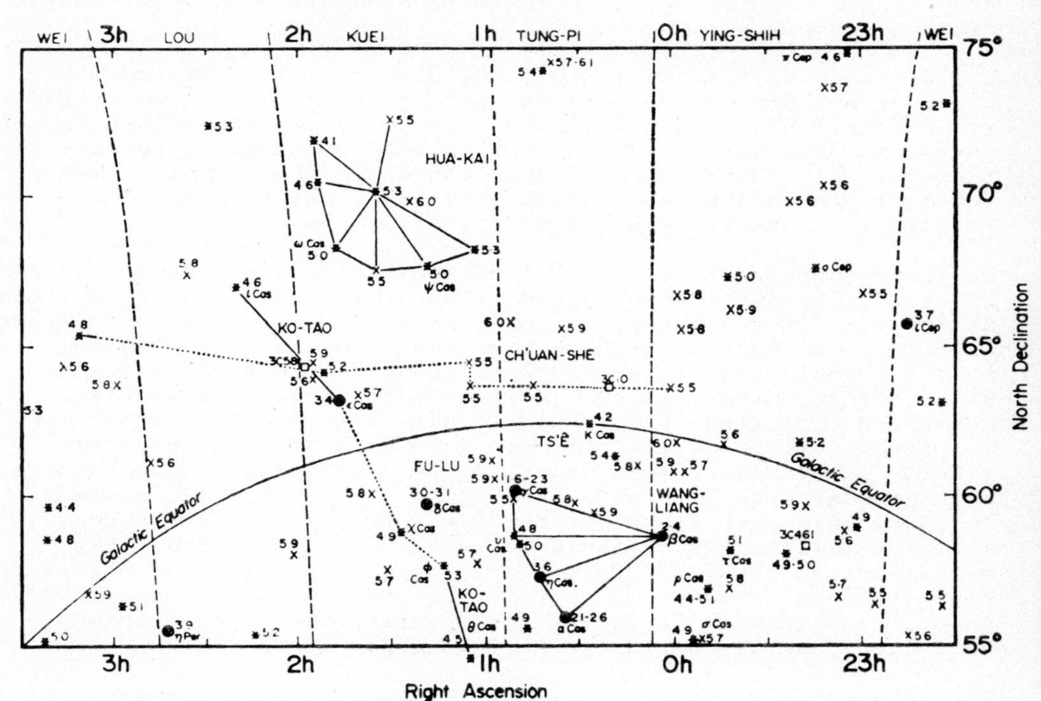

Fig. 9.1. Asterisms in the region of Cassiopeia.

The references to Ch'uan-shê, Hua-kai, and Wang-liang in the Chinese and Japanese accounts of the guest star are quite independent. All of the star maps consulted (see chapter 2) show these constellations, but the form of each differs to a greater or lesser degree from one map to another, The stars forming Hua-kai and Wang-liang can be identified with confidence, for the shape of both of these asterisms is very well defined. However Ch'uan-shê is very extensive and consists of faint stars. In consequence, identification of the constituent stars is more difficult. There are two references to this asterism in the guest star accounts - from southern China (Sung-shih) and Japan (Meigetsuki or Dainihonshi). It is fortunate that the Chinese observation is the more valuable, for three of the star charts used in this investigation were constructed in China at about this time. These star charts were discussed in chapter 2.

It seems that stars brighter than mag.+5 were seldom omitted from the oriental asterisms. Both Hua-kai and Ch'uan-shê certainly contain several stars as faint as +5.5 but none fainter than this. As already mentioned, Hua-kai can be reconstructed reliably. Two of the stars of this group are of magnitude +5.5, while Yú-kuei, the 23rd lunar mansion (the trapezium in Cancer surrounding Praesepe), contains two stars of magnitude +5.5 and +5.6. The limit for a recognisable feature seems to lie very close to magnitude +5.5; stars fainter than this can probably be glimpsed in favouable conditions. The Chinese astronomers, like their western counterparts, counted only seven stars in the Pleiades (Mao, the 18th lunar mansion). These must have been 17 Tau (mag. +3.8), 19(+4.4), 20(+4.0), 23(+4.3), 25 (+3.0), 27(+3.8) and either 28 Tau (+5.2) or 16 Tau (+5.4). Other stars in this cluster brighter than +6.0 are 18 Tau (+5.6), 21 Tau (+5.9) and BD +22° 563 (+5.5). All of these stars must have been overlooked.

As well as the asterisms Ch'uan-shê, Hua-kai and Wang-liang it is necessary to attempt reconstruction of Ko-tao which lies close to all of these formations. On the Su Sung chart, Ko-tao actually intersects Ch'uan-shê, a very rare occurrence. Two isolated stars, Tsê-hsing which forms part of Wang-liang, and Fu-lu, part of Ko-tao, are also included in this investigation.

(a) Hua-kai

According to the Chin-shu (Ho Peng Yoke, 1966), "On top of (the star) Ta-ti (polaris) and suspended above the Imperial Throne are the seven stars of Hua-kai ("Gilded Canopy"). Below (i.e to the north of) Hua-kai forming its handle are the nine stars of K'ang ("Flagstaff")." The shape of this compact group is so characteristic that although the constituent stars are rather faint, it is immediately recognizable on all the star maps we consulted. The formation is most carefully portrayed on the Jesuit, Su Sung and Soochow maps. The Jesuit maps show Hua-kai lying in K'uei-hsiu. On both the Soochow and Su Sung maps the asterism is situated in Lou, the adjacent lunar mansion, but this is obviously an error. The stars are without doubt 32 Cas (mag.+5.3), 36 ψ (+5.0), 40 (+5.5), 43 (+5.5), 46 (+5.0), 48 (+4.6) and 50 Cas (+4.1). The star 40 Cas, at the north, probably forms part of the "flagstaff" (K'ang), which extends much to the north.

(b) *Wang-liang and T'sê-hsing*
Named after a famous early charioteer, this group represents a chariot and a team of four horses, together with a whip. By definition, "*Wang-liang*, comprising five stars north of *K'uei* (15th lunar mansion) lies at the centre of the Milky Way (i.e at its northernmost part) and (denotes) the charioteer of the Son of Heaven. Four of these stars are known as *T'ien-szŭ* ("Celestial Quadriga") while the one by the side is called *Wang-liang* .... The single star called *T'sê-hsing* ("Whip Star"), the whip used by the *Wang-liang*.....is found by the side of *Wang-liang*" (Ho Peng Yoke, 1966). On all the maps we used β Cas represents *Wang-liang* himself. The main charts seem agreed that the remaining stars (representing theteam & whip) are(in order of R.A), κ , α , η , ν (regarding ν^1 and ν^2 as one star) and γ Cas. Whereas on the Jesuit planispheres γ Cas is named as *T'sê-hsing*, the other charts point to κ Cas at the north. Support for κ Cas as the whip comes from the records of the supernova of AD 1572. Both Chinese and Korean records describe this star as being *T'sê-hsing*. The corresponding radio source (3C 10) is only 1.2 degrees north-west of κ Cas. The distance from γ Cas is 5 degrees. We have adopted κ Cas as the "whip".

(c) *Ko-tao and Fu-lu*
These asterisms are described in the *Chin-shu* as follows. "*Ko-tao* ("Hanging Gallery"), consisting of six stars in front of *Wang-liang*, like a flying path (across mountains), stretches from the "Purple Palace" (the circle of perpetual visibility) to the Milky Way.... The star *Fu-lu* ("Auxiliary Road") south of *Ko-tao* (represents) the alternative side-routes."

Both *Ko-tao* and *Fu-lu* show considerable diversity of form on the various charts. On the Jesuit maps the stars of *Ko-tao* are certainly ι , ε , χ , φ , θ , and either ξ or ν Cas. *Fu-lu* is η Cas. These last three stars lie to the south of the region covered by Fig. 9.1. It is clear from the relative positions of ι Cas, ε Cas and BD +63°265 (the +5.2 mag. star lying between ι and ε Cas) that the latter star forms part of *Ch'uan-shê*. On the Su Sung map the *five* stars of *Ko-tao* are ι, ε , χ , φ and θ Cas, while *Fu-lu* is δ Cas. Here *Ch'uan-shê* intersects *Ko-tao* between ι and ε Cas. *Ch'uan-shê* is very idealized, but it is clear from the position of the *K'uei-Lou* boundary that BD +63°265 is one of the stars of this asterism. The Korean planisphere appears to confirm the Jesuit identification of *Ko-tao* but replaces η Cas by χ Cas as *Fu-lu*.

The remaining representations are very idealized. The Soochow chart points to δ Cas as *Fu-lu*, but *Ko-tao* is very distorted, most probably containing θ , φ, χ and ε Cas. On the Tunhuang map *Ko-tao* and *Fu-lu* form a single asterism, presumably, from its shape, consisting of ι , ε, χ, δ , φ and θ Cas.

The formation shown in Fig. 9.1 seems to be the most probable one, but obviously there is considerable doubt about *Fu-lu*. However, we can be fairly confident that within the boundary of Fig. 9.1 all of the stars ι , ε , χ , δ, φ and θ Cas, and only these, are members of the two asterisms.

(d) *Ch'uan-shê*

From the *Chin-shu*; "*Ch'uan-shê* ("Inns" or "Guest Houses"), comprising nine stars and situated above (i.e to the south of) *Hua-kai* near the Milky Way, represents the guest-houses......" (Ho Peng Yoke, 1966). This undistinguished constellation is very extensive. All of the maps (except those in the *Hsing-ching*) show *Ch'uan-shê* situated immediately to the south of *Hua-kai*. On the Jesuit, Soochow and Korean maps the asterism is represented as lying to the north of, and entirely outside, the Milky Way - in accord with the *Chin-shu* description.

The Jesuit representations of *Ch'uan-shê* show only five stars. Measurement of position indicates that four of the stars are 10 Cas (mag. +5.5), BD +63°99 (+5.5), BD +63°149 (+5.5) and BD +63°265(+5.2). The fifth is probably 32 Cas (+5.5) displaced so that it is in line with the others. This formation is shown in Fig. 9.1. On both the Soochow and Su Sung maps *Ch'uan-shê* ranges much farther to the east, but the five easternmost stars appear to be identical with those on the Jesuit planispheres. The gap between BD 63°265 and the first of the eastern stars, probably BD 65°340 (mag.+4.8), is clearly shown in each case. On the Soochow chart the group has a zigzag pattern, seemingly an attempt to portray the true configuration.

All of these stars are faint but it is difficult to accept alternative identifications; there are no other suitable stars between *Hua-kai* and the Milky Way. The star κ Cas, identified above as *T'sê-hsing* is certainly not part of *Ch'uan-shê* for it lies well within the northern edge of the Milky Way. The occurrence of two independent references to *Ch'uan-shê* in the accounts of the guest star suggest that it was a well-known formation. This is apparent from the frequent allusions to the asterism in cometary records (e.g in AD 451, 626, 1337 and 1593).

The expression *fan* ("invade", "offend", "trespass against", etc.) which occurs in the *Sung-shih* account of the guest star is very frequently used in far eastern records to describe conjunctions of the Moon and planets with one another and with stars. The term is defined in chapter 157 of the *Wu-pei-chih* ("Treatise on Armament Technology" - dated AD 1628) in several ways. These are mainly descriptive, but a numerical definition is as follows: "When (a celestial body) comes within 7 *ts'un* of (another) such that their rays extend towards each other" (Ho Peng Yoke, 1966). The *Chin-shu* underlines the astrological significance of the expression: "When the planets are within a distance of seven inches (*ts'un*) from each other the omen will assuredly be fulfilled." This apparent confusion of linear and angular units was discussed in chapter 8 where it was concluded that *ch'ih* (of 10 *ts'un*) and *tu* (degrees, numbering 365.25 to a circle) are synonymous or at least roughly equivalent.

An investigation of numerous Chinese and Korean observations of the Moon and planets invading one another, or stars, reveals that *fan* was usually used when the least separation was about half a degree, and only very rarely for a minimum approach of more than a degree. This seems to confirm the figure of 0.7 degrees from the Wu-pei-chih. When an occultation occurred the more precise terms yen (conceal), shih (eclipse), or ju (enter) were used. Occasionally very close approaches of planets to one another, or to stars, were described in these ways. An unexpected outcome of the investigation is the evidence that although the civil day began at midnight, the astronomers of both countries adopted an astronomical day beginning at sunrise - normally a readily defined moment.

The use of the expression fan does not necessarily imply motion. In the context of the guest star record the reference is to a star, not previously seen, suddenly appearing close to one of the permanent members of an asterism. In order to examine the reliability of the term as it applied to Sung observations we have analysed a series of planetary conjunctions which were observed between AD 1167 and 1200. These are recorded in the Sung-shih and cover four reign periods. In each case it is stated that one planet invaded another. The result of this analysis are shown in Table 9.1.

Ref. No.	Recorded Date	Planets	Actual Date	Time	Distance	Elong.
1	1167 Dec. 24	V,J	Dec. 24	20^h	0.57°	53°E
2	1168 Apr. 14	M,S	Apr. 18	2	0.65	78 W
3	1170 Sep. 9	M,J	Sep. 13	5	0.09	106 W
4	1172 Jun. 18	V,J	Jun. 18	20	0.97	17 W
5	1173 May 11	M,J	May 12	21	0.61	71 E
6	1173 Sep. 5	V,J	Sep. 6	5	0.32	18 W
7	1175 Oct. 25	V,M	Oct. 25	18	0.41	32 E
8	1181 Aug. 14	V,S	Aug. 15	4	0.16	23 W
9	1184 Aug. 31	V,J	Sep. 1	4	0.69	35 W
10	1187 Nov. 14	V,S	Nov. 15	6	0.47	36 W
11	1189 May 22	V,M	Jun. 2	4	1.52	40 W
12	1191 Dec. 31	V,J	Dec. 31	18	0.38	45 E
13	1195 Oct. 11	V,M	Oct. 12	5	0.30	41 W
14	1198 Nov. 18	V,J	Nov. 19	6	1.06	46 W
15	1199 Dec. 2	M,J	Dec. 5	6	0.56	34 W
16	1200 Jan. 1	V,S	Jan. 1	18	0.95	45 E
17	1200 May 22	M,S	May 25	4	0.26	83 W

Table 9.1.
Planetary conjunctions recorded in the Sung-shih.

Column 2 of Table 9.1 gives the recorded date converted to the Julian Calendar using the tables of Hsüeh Chung-san and Ou-yang I (1956). Column 3 gives the initial letter of one of the four planets Venus, Mars, Jupiter, and Saturn (observations of Mercury are extremely rare). Columns 4 and 5 give the calculated date and local time (to the nearest hour) of closest visible approach. The time chosen is the most favourable time for observation of the conjunction, and is within 12 hours of the time of closest approach. Positions of Venus and Mars were computed using the orbital elements given by Newcomb (1895). For each planet an unperturbed orbit was assumed. This treatment is quite adequate for the present purpose. Positions of the slower moving outer planets were found by interpolating in Tuckerman's Tables (1964). Column 6 gives the separation of the two planets at the time given in column 5. Column 7 gives the elongation of the planets east or west of the Sun.

Of the 17 observations, the recorded dates of 11 are exact (1,4,6, 7,8,9,10,12,13,14 and 16). For observation No. 11 the date is 10 days in error. This implies that the cyclical day i-wei of the record should be replaced by i-szŭ although the characters wei and szŭ are quite dissimilar. This observation seems to be very unreliable, for the closest visible approach (1.52 degrees) is about 0.5 degrees larger than any other in Table 9.1. For the remaining 5 observations (2,3,4,15 and 17) the Sung-shih records a date differing by up to 3 days from the calculated date. If the record gives the true date of observation, the amended separations (in degrees) are respectively 1.31,0.88,0.78,1.06 and 1.35. It seems quite possible that cloud interfered with some of the observations. However, the average of these five results is about 2.5 times that of the corresponding figures in Table 9.1 and twice that of the remaining quantities in the table (excluding No. 11). An error in the recorded date is thus a more probable assumption. This is a common occurrence in lunar records, for which the intended date can be established with certainty owing to the rapid motion of the Moon. The Sung planetary observations thus indicate a maximum separation of about a degree for the term fan to be applicable, in accord with the limit deduced from other oriental observations.

As discussed in chapter 11, an interesting Korean observation of the planet Venus invading the guest star of AD 1604 (Kepler's supernova) is recorded in the Sŏnjo Sillok on a date which corresponds to AD 1605 January 20 (Gregorian). Calculation shows that before dawn on January 21 (astronomical date January 20) in AD 1605, Venus and the star would be seen only 0.51 degrees apart. It thus seems probable that throughout the far east a sighting tube was used in order to decide when the expression fan was appropriate.

Within the right ascension range 22^h30^m to 3^h30^m, and north of declination $+55^o$, the Clark and Caswell (1976) catalogue lists six SNRs; Cas A, CTB1, CTA1, 3C 10, 3C 58, and HB3. CTB1, CTA1, and HB3 are all of large angular size and very low surface brightness indicative of their being of extreme age. 3C 10 is the well-established remnant of Tycho's supernova (chapter 10). As already mentioned in chapter 4, Cas A is undoubtedly the remnant of a recent (≃200 years), nearby, but undetected, supernova. 3C 58 thus remains as the only "young" SNR in the region of interest which could be the remnant of the AD 1181 event.

The SNR 3C 58 fits in a general sense the various oriental descriptions of the position of the guest star of AD 1181 - "in K'uei-hsiu and invading the stars of Ch'uan-shê" (Sung-shih): "at Hua-kai" (Kin-shih): "near Wang-liang and guarding Ch'uan-shê" (Meigetsuki and Dainihonshi). As shown in Fig. 9.1, 3C 58 lies just within K'uei-hsiu. It is 8 degrees from Wang-liang but this constellation contains some of the brightest stars in Cassiopeia. The proximity to Hua-kai (4 degrees) is in accord with the Kin account.

The mention of the expression fan ("invading") in the Sung-shih record is critical. As already seen this is an important astrological term implying a maximum distance of about one degree from a celestial body. 3C 58 lies only 0.68 degrees from BD 63°265 (mag.+5.2) and 0.53 degrees from 53 Cas (mag.+5.6) in Ch'uan-shê (the "Guest Houses").

Bearing in mind the ample opportunity for observation of the guest star, 3C 58 fits all of the evidence from the purely astronomical point of view. The only difficulty in accepting this identification is the lack of reference to Ko-tao. However, this asterism is only rarely mentioned in accounts of both comets and novae (in contrast to Ch'uan-shê). Possibly the explanation is astrological.

The true nature of the source 3C 58 has been the subject of considerable debate in recent years. On the basis of its angular extent, low galactic latitude, and flat spectrum (α =-0.1) it was originally classified as an H II region. Its non-thermal nature was eventually revealed from polarisation measurements by Weiler and Seielstad (1971), and Kundu and Velusamy (1972). There can now be no doubting a synchrotron origin for the continuum radio emission from the source. Nevertheless high-resolution maps (see Fig. 9.2) fail to show the peripheral brightening characteristic of a supernova remnant. The amorphous structure of the source and unusually flat spectrum immediately liken 3C 58 to the Crab Nebula radio source; here the similarity ends, since no central pulsar or optical nebulosity has yet been detected for 3C 58.

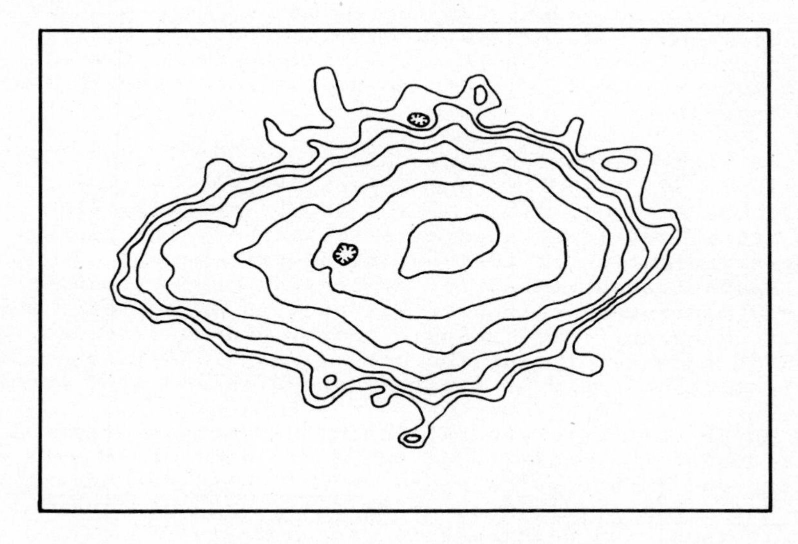

Fig. 9.2. A 15 GHz map of 3C 58 (from Green et al, 1975)

H I absorption measurements by Goss et al (1973), and Williams (1973) place 3C 58 at or beyond the distance of the outer galactic spiral arm at about 8 kpc. Despite this distance, the large z-value for the source ($\sim$150 pc) and consequent lower-than-average interstellar absorption would still allow an observed magnitude for the supernova of about 0 (see earlier). Since the H I absorption results represent a minimum distance, there remains the possibility that 3C 58 could be extragalactic; indeed there are sources with similar amorphous structure (for example 3C 386) which have been positively identified as being extragalactic. However the flat spectrum of 3C 58 is even less characteristic of an extragalactic source than of an SNR. Despite these misgivings, the majority of investigators now favour an SNR interpretation for the source.

The AD 1181 supernova presents another example where the possible association of an SNR with the new star depends critically on the exact interpretation of a single word in the historical records. In this case we must rely on the astronomical precision which we have argued is implied by the use of *fan* (invading); then the excellent positional agreement between the radio source 3C 58 (R.A $2^h01^m43^s$, dec.+$64^o36.5'$) and the star BD 63^o265 (R.A $1^h55^m55^s$, dec.+$64^o22.8'$, with negligible proper motion) in *Ch'uan-shê* is convincing evidence for 3C 58 being the SNR. In fact there is no other remnant in the vicinity that could possibly correspond, although because of some remaining doubts as to the true nature of the radio source we feel we must limit the classification in this case to 3C 58 being the *probable* remnant of the supernova of AD 1181.

The Sung dynasty astronomers hold the distinction of having recorded three supernovae - those of AD 1006, 1054, and 1181. It was to be almost 4 centuries before another supernova was sighted from Earth, by which time the standard of European observational astronomy had far surpassed that of the Far East.

Chapter 10

THE NEW STAR OF TYCHO BRAHE

During the almost four centuries which were to elapse before the next supernova appeared, very little progress was made in Far Eastern astronomy. The new star which appeared in AD 1572 was observed both in China and Korea (there is no known extant record from Japan), but the reports from these countries are in no way superior to those of new stars seen a thousand years earlier. However, in Europe the Renaissance was now well under way. We today are indeed fortunate that an observer of the calibre of Tycho Brahe was flourishing when the star appeared. The careful records which this eminent Danish astronomer (who lived between AD 1546 and 1601, and was not quite 26 when the new star appeared) has left us prove beyond any reasonable doubt that the star was a supernova of Type I, and allow us to establish with certainty that its remnant is the powerful non-thermal galactic radio source G120.1+1.4 (3C10).

The star appeared at a very critical time in the development of astronomical thought in Europe. Copernicus' heliocentric theory of the Universe, published in his _De Revolutionibus Orbium Coelestium_ (AD 1543), was beginning to attract widespread attention. In this work Copernicus attacked the long-held doctrine of Aristotle and Ptolemy that the Earth was at rest at the centre of the Universe. This had formed the basis of medieval European cosmology. It was one of the basic tenets of Aristotle's concept of the Universe that the eighth sphere of the fixed stars (the first seven spheres were taken up by the Moon, Sun and five known planets) was immutable. Hence Tycho Brahe, who could not conceal his amazement at the sudden appearance of the new star (which was about as bright as Venus), was at great pains to determine whether it belonged to the eighth sphere. In a preliminary report entitled _De Nova Stella_ and published in AD 1573, Tycho discussed his measurements of the position of the star in relation to the principal stars of Cassiopeia (within which constellation the new star appeared). He concluded: "That it is neither in the orbit of Saturn, however, nor in that of Jupiter, nor in that of Mars, nor in that of any one of the other planets, is hence evident, since after the lapse of several months it has not advanced by its own motion a single minute from that place in which I first saw it; which it must have done if it were in some planetary orbit..... Hence this new star is located neither in the region of the Element, below the Moon, nor in the orbits of the seven wandering stars but in the eighth sphere, among the other fixed stars". Tycho's measurements allow us to determine the position of the new star at the present day with remarkable accuracy - to within a very few minutes of arc.

Tycho's original discussion of the new star is very brief. However, in his Astronomiae Instauratae Progymnasmata, which was published in AD 1602, shortly after his death, he gives a very detailed investigation, including numerous observations made by others. In this work the De Nova Stella is also reprinted, with minor omissions, as part of chapter 8.

Coming now to the star itself, it could have scarcely been better placed for observation in Europe and the Far East in view of its location in Cassiopeia. North of about latitude 35°N, this constellation is circumpolar (i.e it never sets), and in early November, when first discovered, the star would be almost in the zenith in the evening. An excellent and objective account of the discovery in Europe is given by Dreyer (1890). From this it seems certain that the star appeared some time between November 3 and November 6 (in AD 1572). Dreyer rejects a number of spurious claims, several of which are clearly self contradictory, that the star was sighted in October. Most of these are quoted in the Progymnasmata, but perhaps the most fascinating story is contained in a French pamphlet, unknown to Tycho Brahe, and cited by Dreyer. This little work, published in 1590, and entitled La nouvelle Estoille apparue sur tous les Climats du Monde: Et de ses éffects, claims that the star was seen in the month of October by Spanish shepherds "keeping watch over their flocks" (cf St. Luke, ii,8).

A valuable negative sighting in early November is given by Jerome Mugnoz (Muñoz), an eminent mathematician and Professor of Hebrew in the University of Valencia (Spain). The original Spanish is quoted by Hellman (1960). This account runs as follows: "I am certain that on the second day of November 1572 there was not this comet in the sky, because on purpose more than an hour after 6 in the afternoon, I showed in Hontinente (Onteniente) to many people publicly how to identify the stars, and there were many shepherds very experienced in them (the stars), who informed me on the 18th day in the morning a new star had appeared". Mugnoz later measured the position of the new star in relation to three of the stars in Cassiopeia with fairly high precision (see below) so that his testimony is probably reliable. Mugnoz was not alone in describing the new star as a comet . Possibly he was being cautious. His reference to the shepherds is somewhat obscure, for by the 11th of the month, when Tycho Brahe first saw the star, it was already about as bright as Venus. It seems best to trust only the first hand experience of Mugnoz himself.

There does not seem to have been any positive sighting of the new star until November 6, when Wolfgang Schüler of Wittenberg discovered it at about 6 a.m (Progymnasmata, chapter 9). Like Mugnoz, Schuler described it as a comet. There is the possibility that Francesco Maurolyco of Messina may have preceded Schuler. Hellman (1960) has discussed in detail a manuscript copy of Maurolyco's account of the star, which states that "Maurolycus, abbot of Messina, wrote this hastily, subject to the judgement of the more wise. 6 November 1572". This suggests that Maurolyco may have seen the star earlier than November 6.

By November 7 the star had been sighted independently by two others.Tycho Brahe himself did not see it until the evening of November 11. Bad weather had delayed the discovery of the star by this assiduous observer. He tells us: "Although several people had preceded me (in seeing it), the air in our part of the world had not been clear enough" (Prog., chapter 3).

Valuable supporting evidence for a date of first appearance in early November is given by the court astronomers of China. The star was first seen in China on a date corresponding to November 8. Unfortunately, only the month of discovery is reported in Korea, but this corresponds to the period November 6 to December 4.

While on the subject of the discovery of the new star, it is worthwhile commenting on the attitude to it in Europe. In chapter 3 of his Progymnasmata, Tycho Brahe gives a graphic account of his own feelings. When the star first appeared, Tycho was staying with his uncle at the monastery of Herritzwadt in Denmark. He writes :
"When on the above mentioned day (November 11), a little before dinner.....I was returning to that house, and during my walk contemplating the sky here and there since the clearer sky seemed to be just what could be wished for in order to continue observations after dinner, behold, directly overhead, a certain strange star was suddenly seen, flashing its light with a radiant gleam and it struck my eyes. Amazed, and as if astonished and stupified, I stood still, gazing for a certain length of time with my eyes fixed intently upon it and noticing that same star placed close to the stars which antiquity attributed to Cassiopeia. When I had satisfied myself that no star of that kind had ever shone forth before, I was led into such perplexity by the unbelievability of the thing that I began to doubt the faith of my own eyes, and so, turning to the servants who were accompanying me, I asked them whether they too could see a certain extremely bright star when I pointed out the place directly overhead. They immediately replied with one voice that they saw it completely and that it was extremely bright. But despite their affirmation, still being doubtful on account of the novelty of the thing, I enquired of some country people who by chance were travelling past in carriages whether they could see a certain star in the height. Indeed these people shouted out that they saw that huge star, which had never been noticed so high up. And at length, having confirmed that my vision was not deceiving me, but in fact that an unusual star existed there, beyond all type, and marvelling that the sky had brought forth a certain new phenomenon to be compared with the other stars, immediately I got ready my instrument. I began to measure its situation and distance from the neighbouring stars of Cassiopeia, and to note extremely diligently those things which were visible to the eye concerning its apparent size, form, colour and other aspects".

The general reaction in Europe to the appearance of the star was one of extreme conservatism. Michael Maestlin, later to become tutor to Johannes Kepler, remarked: "Various judgements and opinions were heard about it from people who are not illiterate.

Some conceded that it was natural and permanent, not new....... Others indeed believed that it was Vega, others Capella, others Arcturus ; but others (said) Venus, others Saturn, others some other star or planet" (Prog., chapter 8). The well known astronomer Cornelius Gemma did not begin to observe the position of the star until November 26, since he thought it idle talk when he first heard of a new star (Prog., chapter 8). Tycho himself wrote that he later heard that in Germany country people first called attention of astronomers to the phenomenon (Prog., chapter 3).

There could well be truth in this last remark, for unlettered people would be relatively free from preconceived ideas. Certainly if a star as bright as Venus could be overlooked by much of the astronomical world as late as AD 1572, there should be no difficulty in understanding the complete lack of European references to the supernova of AD 1054. Here, the vast majority of works which could have mentioned the star are monastic chronicles, the writers of which might be expected to hold very orthodox views.

Before discussing the European measurements of the position and brightness of the star, let us take a look at the Far Eastern observations.

The most detailed account is to be found in the Ming-shih-lu, in the annals of Emperor Shên-tsung (the Shên-tsung-shih-lu). We read: "(6th year of the Lung-ch'ing reign period, 10th month), (day) hsin-wei.....Previously in the 10th month on the 3rd day ping-ch'en at night a guest star was seen at the north-east; it was like a crossbow pellet. It appeared beside Ko-tao in the degrees of (Tung-) pi lunar mansion. It gradually became fainter. It emitted light in the form of pointed rays. After the 19th day jen-shen at night the said star was orange (reddish-yellow) in colour. It was as large as a lamp and the pointed rays of light came out in all directions........It was seen before sunset...... At the time the Emperor saw it in his palace. He was alarmed and afraid, and at night he prayed in the open air on the Vermillion Steps......Editorial note: The star gradually diminished in brightness only in the 2nd month of the 1st year of the Wan-li reign period. When we come to the 4th month of the 2nd year it finally disappeared".

The first part of the above account is retrospective, for jen-shen is the day after hsin-wei, the date of the previous entry. Evidently the writer of the above account was more interested in the Emperor's reaction (presumably he first saw the star on the day jen-shen). We have omitted two interpretations of the star in the text since they disturb the flow of the account. The first states that the star was a "rayed star" (po-hsing), and second that such stars "are also seen in the daytime" in agreement with the observation.

The date of discovery of the new star corresponds to AD 1572 November 8, and the subsequent date when the colour was orange and the Emperor saw the star is equivalent to AD 1572 November

24. The month when the star showed a noticeable diminution in brightness corresponds to AD 1573 March 1 to 31, and the month of final disappearance to AD 1574 April 21 to May 19.

In the astronomical treatise of the Ming-shih, the obvious place to expect a detailed account of the star, we find only a brief undated mention (chapter 25): "There are also some (stars) which did not exist in antiquity but which exist now. Beside Ts'ê-hsing there is a guest star which newly appeared during the first year of the Wan-li reign period. At first it was large, but now it is small".

There is no mention of the star in the Ming-shih annals, but in general we have found the imperial annals of the various dynastic histories to be unproductive as far as astronomical records are concerned. However, Ho Peng Yoke (1962) discovered a brief mention of the star in the Ming-shih-kao, the draft version of the Ming-shih. Here the guest star is described as a 'broom star' (hui-hsing) but this error was corrected in the astronomical treatise of the Ming-shih. We read in the Ming-shih-kao : "(6th year of the Lung-ch'ing reign period), winter, 10th month (day) ping-ch'en. A broom star appeared in the north-east. It disappeared only during the 4th month of the 2nd year of the Wan-li reign period."

The single Korean record is also very brief. In the Sŏnjo Sillok ("Annals of the reign of King Sŏnjo") we find the following account of the star: "(5th year of Sŏnjo) 10th month. A guest star was seen at (yu) Ts'ê-hsing. It was larger than Venus."

If all we had to go on were the oriental records of the star, it would be immediately apparent from the lengthy duration (some 18 months) that the star was not a comet, and for just this same reason a supernova interpretation would be more likely than a nova. Additionally, the Ming-shih and the Sŏnjo Sillok independently mention the proximity of the star to Ts'ê-hsing. As was shown in Chapter 9, this star is identical with κ Cas.. This location in confirmed by a star map produced by Jesuits in Peking around AD 1600, and preserved in Bologna (see chapter 2). Here the star is shown about 1 degree to the west of a star whose position corresponds with that of κ Cas.

However, the positional measurement of Tycho Brahe are fully two orders of magnitude better than those of the oriental astronomers, and further, as Baade showed in 1945, it is possible to draw a remarkably accurate light curve from Tycho's comparisons with nearby stars.

Considering first the light variation of the star, the negative sighting by Mugnoz is quite valuable. Cassiopeia is one of the most familiar of all constellations since the principal stars form a distinct "W". The magnitudes of these stars lie in the range from about +2.5 to +3.5. As is evident from Fig. 10.1, the location of the new star was such that unless it was very faint it could scarcely have failed to attract attention if someone was looking closely at the constellations. We might thus reasonably

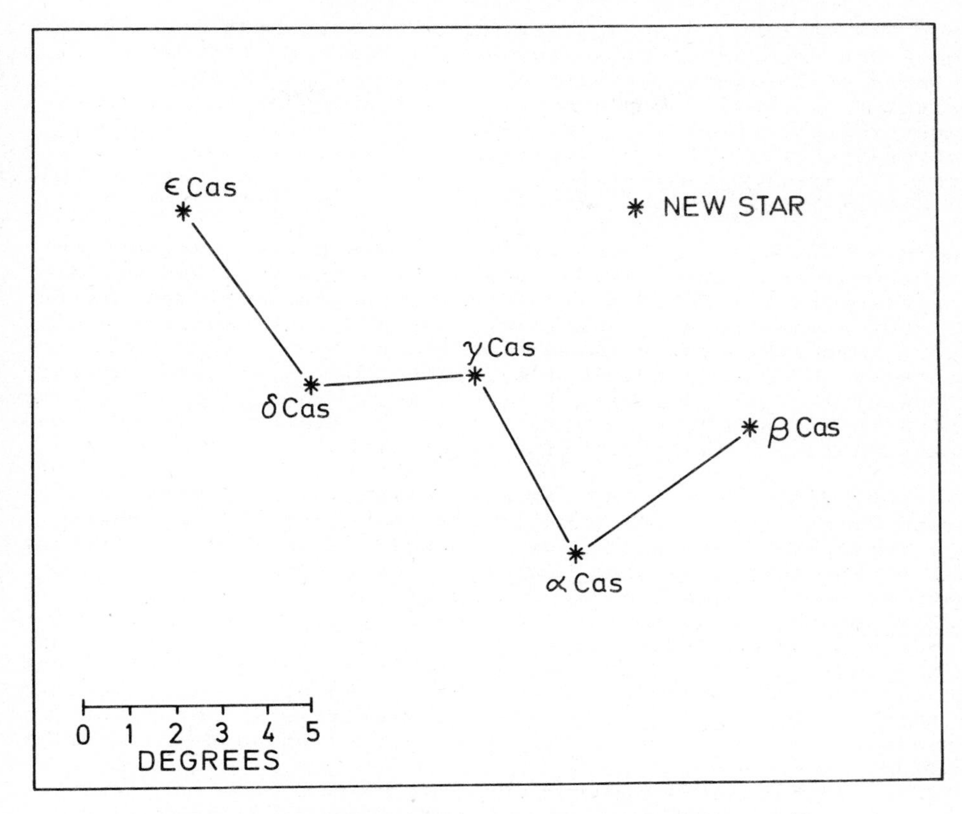

Fig. 10.1. The position of the new star with respect to the "W" of Cassiopeia.

conclude that on the evening of November 2, when Mugnoz was giving his demonstration, the new star had not reached magnitude +3.

Several observers, including Tycho Brahe, made estimates of the brightness of the new star around maximum brilliancy, but unfortunately it is not possible to establish the date of maximum with any accuracy. Tycho (<u>Prog</u>., chapter 3) estimated that when he first saw the star (November 11) it emulated Venus when the planet was at maximum brightness (which was virtually the case then). He added that it remained at nearly this same brightness throughout almost the whole of November. He further remarked that when the sky was clear many keen sighted people could see it in daylight, even at midday, while at night the star often shone through thick clouds which blotted out the other stars. However, several observers felt that the new star never quite rivalled Venus. Baade (1945) in his pioneering investigation of the light curve of the star gave considerable weight to three other estimates made around maximum. The observers were Caspar Peucer and John Praetorius, both of Wittenberg, and Michael Maestlin of Tubingen. Peucer noted that the new star was "brighter than all the planets and star with the exception of Venus" (Prog., chapter 8). According to Praetorius, it was "larger and brighter than Jupiter, but easily fainter than Venus" (Prog., chapter 9). Maestlin estimated that it "surpassed Jupiter, and almost Venus" (Prog., chapter 8).

From Baade (1943), in November of 1572 Venus was near maximum brightness, (magnitude -4.3), and approaching greatest western elongation from the Sun after inferior conjunction. The planet was thus a morning star, and since the new star was circumpolar, direct comparison could be made in the early morning hours. Combining the various estimates, Baade deduced an apparent magnitude of the star at maximum as -4.0 ± 0.3. This is in fair agreement with the Korean observation that the star was "larger than Venus" and the Chinese sighting in daylight.

With regard to the date of maximum, we can do no better than follow Baade in assuming a date near the middle of November. Only Tycho Brahe seems to have continued to estimate the brightness of the star after maximum, but fortunately he did this month by month. His estimates, as given in chapter 3 of the Progymnasmata, are summarised in Table 10.1. It is perhaps to be regretted that Tycho did not express dates more accurately than to the nearest month, but possibly he felt that the decline was so slow that this was justified. What is particularly important is that he does not hint at any irregularities in the light variation - there seems to have been a continuous decline after maximum.

	Date	Brightness Estimate
AD 1572	Dec.	About as bright as Jupiter.
1573	Jan.	A little fainter than Jupiter; much brighter than the stars of 1st mag.
1573	Feb.-Mar.	Equal to brighter stars of 1st mag.
1573	Apr.-May	Equal to stars of 2nd mag.
1573	Jul.-Aug.	3rd mag. Similar to brighter stars of Cassiopeia which are of 3rd mag. (i.e. $\alpha, \beta, \gamma, \delta$ Cas.).
1573	Oct.-Nov.	Equal to stars of 4th mag. In Nov. very similar to κ Cas.
1573	Dec.-1574 Jan.	Scarcely exceeded stars of 5th mag.
1574	Feb.	Reached 6th and faintest mag.
1574	Mar.	So faint that no longer visible.

Table 10.1. Tycho Brahe's estimates of the brightness of the new star after maximum.

It is interesting to note that the Chinese astronomers, who continued observing the star until April - May of AD 1574, were able to follow it for a month longer than Tycho. This seems to prove the diligence of the oriental astronomers as straightforward observers, even if the accuracy of their positional measurements was far inferior to that of the Europeans at this time.

In the above Table 10.1, there are no references to June and September in AD 1573. However, Tycho was observing the star in both months. In each case he tells us that the brightness was declining, but it was not until the following month that the star could be placed in a particular magnitude range. Probably he was making comparisons with stars of a given magnitude listed in the catalogue of Ptolemy's Almagest, and during the two months in question he could not find a suitable comparison object. However, there is no doubt that he maintained a continuous watch on the new star.

For the purpose of reducing Tycho's data, Baade assumed a mean date for each comparison. He was particularly careful to check the precise magnitudes of stars in each of the classes 1 to 6 adopted by Ptolemy, and presumably followed by Tycho. It would be difficult to improve on Baade's investigation. His results for the variation in brightness of the new star are summarised in Table 10.2. A date of maximum of AD 1572 November 15 is assumed.

Days after maximum	Visual magnitude
0	-4.0
30	-2.4
61	-1.4
107	+0.3
167	+1.6
259	+2.5
351	+4.0
365	+4.2
412	+4.7
457	+5.3
485	....

Table 10.2. The varying brightness of the new star of AD 1572.

The magnitudes derived by Baade and quoted in Table 10.2 are on the Harvard Revised Photometric scale. Obviously, as the principal objective is to produce a light curve, the precise magnitude scale is unimportant. To the above results may be added the estimate made earlier from the negative sighting by Mugnoz that the star was probably fainter than +3 on November 2 (13 days before the assumed maximum).

The light curve drawn from the data in Table 10.2 is shown in Fig. 10.2. This is taken direct from Baade's paper with the

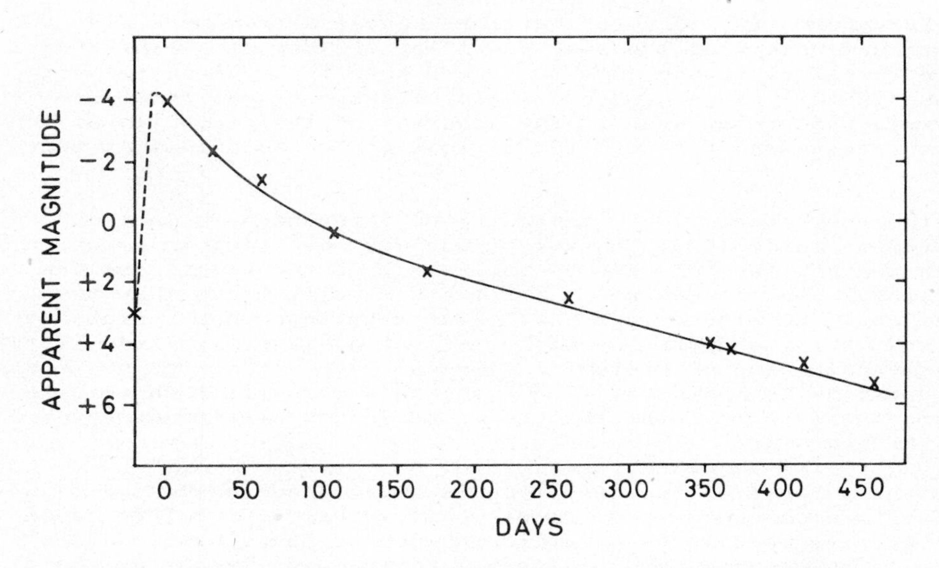

Fig. 10.2. The light curve of the AD 1572 supernova.

addition of the single pre-maximum point. Comparisons with the composite Type I and Type II light curves shown in Figs. 1.6 and 1.7 (chapter 1) make it evident that the star was a supernova of Type I. The light curve is clearly not that of a nova. Undoubtedly Tycho's estimates individually were crude by modern standards, but the resultant light curve has all the characteristics of a Type I supernova - an initial fairly rapid decline followed by a linear descent at about 1 mag. in 50 days.

As the light curve of the new star presents conclusive evidence that it was a supernova rather than a nova, we should expect to find a powerful discrete radio source as its remnant. Tycho Brahe made a far more systematic determination of the location of the star than any of his rivals, measuring its angular distance from the principal stars of Cassiopeia and also Polaris to the nearest minute of arc. From Tycho's measurements, Argelander (1864) and Böhme (1937) deduced position which were in excellent accord. Argelander used Tycho's measurements in relation to $\alpha, \beta, \gamma, \delta, \varepsilon, \zeta, \eta, \kappa$ Cas and obtained a result for the epoch 1950.0 of R.A. $0^h22^m0.2^s$, dec. $+63^\circ52'12''$. Böhme derived two solutions, depending upon whether he excluded η Cas or included it along with ι Cas. The first of these (for the same epoch) is R.A. $0^h22^m1.1^s$, dec. $+63^\circ52'1''$, and the second $0^h22^m4.6^s$, $63^\circ52'30''$. Baade attributed the "unexpectedly large" difference between Böhme's two solutions to an error in Tycho's measurement of the distance of the new star from ι Cas. By omitting this measurement he obtained a result identical with that of Argelander, and estimated the mean error in the position of the star as only $\pm23''$. Baade made a search of the area down to mag. +19 for a blue star which might be the stellar remnant, but found nothing.

Radio emission from the remnant of Tycho's new star was first reported by Hanbury-Brown and Hazard (1952), using a large meridian-transit radiotelescope at the Jodrell Bank Observatory. This discovery, unlike Bolton's chance discovery of radio emission from the Crab Nebula (see chapter 8), was the result of a planned search and represented only the second positive identification of a galactic supernova remnant. The coordinates of the radio source were given by Hanbury-Brown and Hazard as R.A. (1950) $0^h21^m49^s \pm 2^s$, dec.(1950) $+64^\circ15' \pm 35'$, although verification of the discovery in the second Cambridge Catalogue gave a position of R.A.(1950) $0^h22^m46^s \pm 5^s$, dec.(1950) $+63^\circ57' \pm 5'$. On the basis of a further improvement of position in the third Cambridge Catalogue with a correction applied by Fomalont et al (1974), and recent high-resolution maps of the source, the centroid of radio emission is now taken as R.A.(1950) $0^h22^m37^s$, dec.(1950) $+63^\circ52'40''$, and the source diameter as 7.9 arc min. Although the declination agrees exactly with Tycho's measurement of the supernova position, there has been a great deal of speculation and comment on the error of almost 4 arc min in Tycho's right ascension result (for a useful discussion see Minkowski, 1968).

No. (Ptolemy)	Identification	Distance
1	ζ Cas.	10°22'
2	α Cas.	7 50½
3	η Cas.	6 53
4	γ Cas.	5 02
5	δ Cas.	8 03½
6	ε Cas.	9 48
7	ι Cas.	12 58½
11	κ Cas.	1 31
12	β Cas.	5 19

Table 10.3. Tycho's measurements of the angular distance of the new star from some of the principal stars of Cassiopeia.

Today no one would doubt the identification of the radio source (3C10:G120.1+1.4) as the remnant of Tycho's supernova, but the problem of the discrepancy in the position still remains. We have investigated the situation in detail in the hope of obtaining a satisfactory solution. Let us start with the observations themselves.

When the new star appeared, Tycho Brahe had recently finished making a new sextant graduated in minutes of arc. With this instrument he proceeded to measure with extreme care over a period of several months the angular separation of the star from nine of the principal stars of Cassiopeia. His final results are given in chapter 4 of the Progymnasmata in a table entitled: "Distances of the new star from some of the principal fixed stars in the constellation of Cassiopeia, diligently investigated and verified". The essentials of this table are duplicated in Table 10.3. Tycho gave the number and name of each star according to Ptolemy's catalogue. We have identified the stars in modern notation - by their Greek letter.

Tycho also measured the distance of the new star form the Pole Star, and he obtained the result 25°14'. Unfortunately virtually none of his original measurements have survived, although we do know that he had to make large corrections to his results for parallax due to the use of a primitive form of sighting. Examples are quoted by Dreyer (1890): the correction for a measured angle of 5° was 8' and for an angle of 8° was 13', a roughly linear relationship. However, we have no way of testing for systematic errors. In later years Tycho made a series of measurements of the distances of all the 9 reference stars in Cassiopeia from one another. His results are given in chapter 4 of the Progymnasmata, but he tells us that only a small(and unspecified proportion) of the results were obtained with the original instrument.

It is still of interest to examine the general accuracy of Tycho's measurements. We have calculated the distances of the 9 reference stars from one another - a total of 36 results. The mean error in Tycho's figures was 1.1', a remarkable achievement. Errors of greater than 2' were rare, although there was one discrepancy as large as 4.5'.

Our analysis of Tycho's measurements of the distance of the new star from Polaris and the stars of Cassiopeia is shown in Fig. 10.3. Proper motions have been allowed for, but since the observations were made at various (and unknown) times, no corrections could be made for differential refraction, etc. Provided the altitudes were reasonably high the effects of refraction would be negligible. In the diagram, each line represents the equation

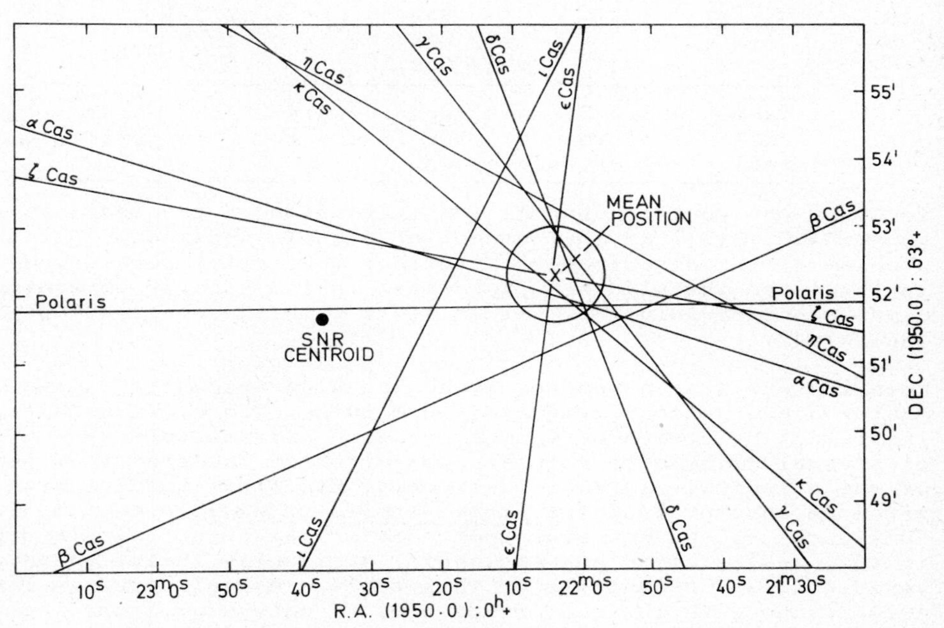

Fig. 10.3. The position of the new star using Tycho's measurements.

between the R.A. and dec. of the new star deduced from Tycho's measured distances. Over the small angular extent of the figure (only a very few minutes of arc) the various lines may be regarded as accurately straight. Ideally, all the lines would be expected to meet in a single point, which would be the location of the new star. This, of course, is not the case in practice owing to random errors of observation and systematic observational and instrumental errors. The small cross near the centre of the figure represents the best least squares fit to Tycho's data, and the circle round it the corresponding standard deviation in the position. It is obvious that the mutual consistency of the individual results is very good, but the final position (R.A. $0^h22^m3.9^s$; dec.$+63^o52'17''$ at epoch 1950.0, with a standard error in both coordinates of $\pm39''$), although in excellent agreement with the previous determination cited above, is some 4 arc min to the west of the centroid of the supernova remnant (shown by a small shaded circle). There must be some systematic error present since proper motion of the SNR is negligible.

If we examine the distribution of the reference stars in relation to the supernova (see Fig. 10.4), we can see that apart from Polaris, all the stars lie within an angle of less than 180 degrees (actually 140 degrees). It so happens that there are no

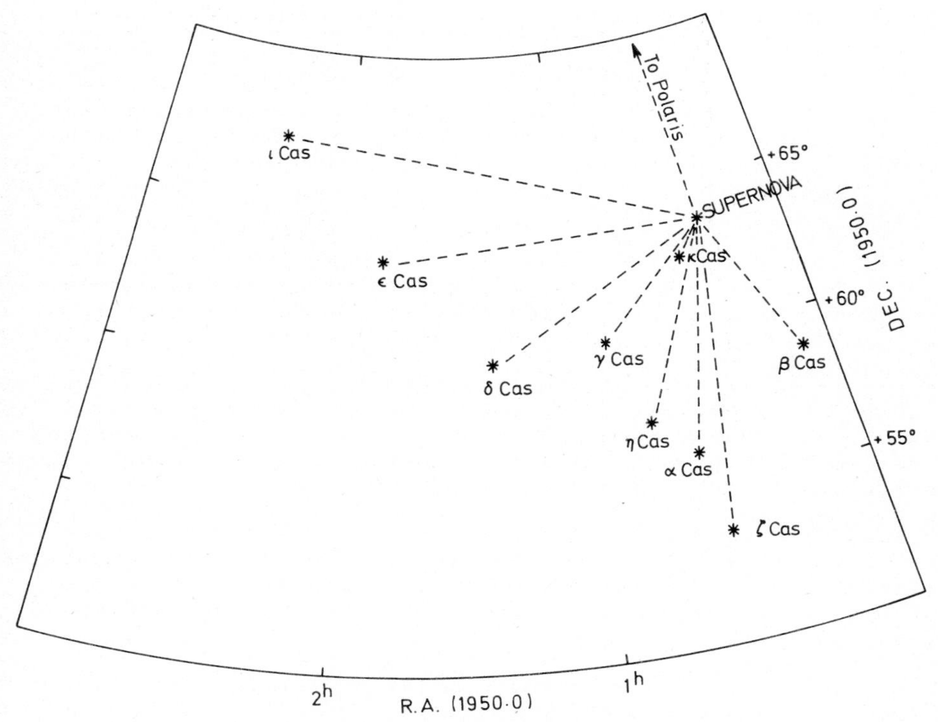

Fig. 10.4. The reference stars used by Tycho in positioning the new star.

nearby stars to the north and west of the position of the supernova, so that any systematic errors would not be expected to reveal themselves. Is it possible to deduce an improved location based on the knowledge that there is a significant discord between the measured position of the new star and the centroid of the radio source? To test this, we calculated the distance between this centroid and the nine reference stars of Cassiopeia (omitting Polaris since the separation is nearly twice any other value) and compared these figures with Tycho's own measurements. The results of this investigation are summarised in Table 10.4.

Star	Distance Calculated	Distance Tycho's	Diff.(C-T)
α Cas.	7°49.1'	7°50.5'	-1.4'
β Cas.	5 20.8	5 19.0	+1.8
γ Cas.	4 58.4	5 02.0	-3.6
δ Cas.	7 59.6	8 03.5	-3.9
ε Cas.	9 44.5	9 48.0	-3.5
ζ Cas.	10 20.7	10 22.0	-1.3
η Cas.	6 49.9	6 53.0	-3.1
ι Cas.	12 57.1	12 58.5	-1.4
κ Cas.	1 28.3	1 31.0	-2.7

Table 10.4. Comparison between the calculated distance of the SNR from the reference stars of Cassiopeia with Tycho's measurements.

The mean of all the differences in Table 10.4 is -2.1'. Apart from β Cas, all the deviations are within about 1 arc min of the mean. The result for β Cas is clearly discordant, although no more so than already noted above for an other measurement. The mean of all the other deviations, which seem to be independent of the angular separation, is -2.6'. Fig. 10.5 is essentially similar to Fig. 10.3 above, but a correction of -2.6' is applied to all Tycho's measurements (with the exception of β Cas and Polaris, which are discarded). The new mean position is once more shown as a cross, and the standard error as the circle surrounding this. The position deduced for the new star is R.A. $0^h22^m32.3^s$, dec.+63°51'17" with a standard error in both coordinates of +56".
In Fig. 10.5, the SNR lies well within the circle of uncertainty, so that the single assumption of a systematic error is justified. This might possibly have arisen through a very slight warping of the wood of the sextant after construction or through Tycho's observational procedure in aligning the edges of the sights on the new star and the comparison star. Whatever the explanation, it should be remembered just how small an error of 2.6' is when observing with the unaided eye. Indeed, even the uncorrected position derived directly from Tycho's data lies within the boundary of the discrete radio source which is the remnant of the supernova.

Virtually nothing has been said so far regarding the measurements of position by Tycho's contemporaries. Many of these results are very crude being up to half a degree in error or even more. Others are reasonably accurate, but this may be merely fortuitous.

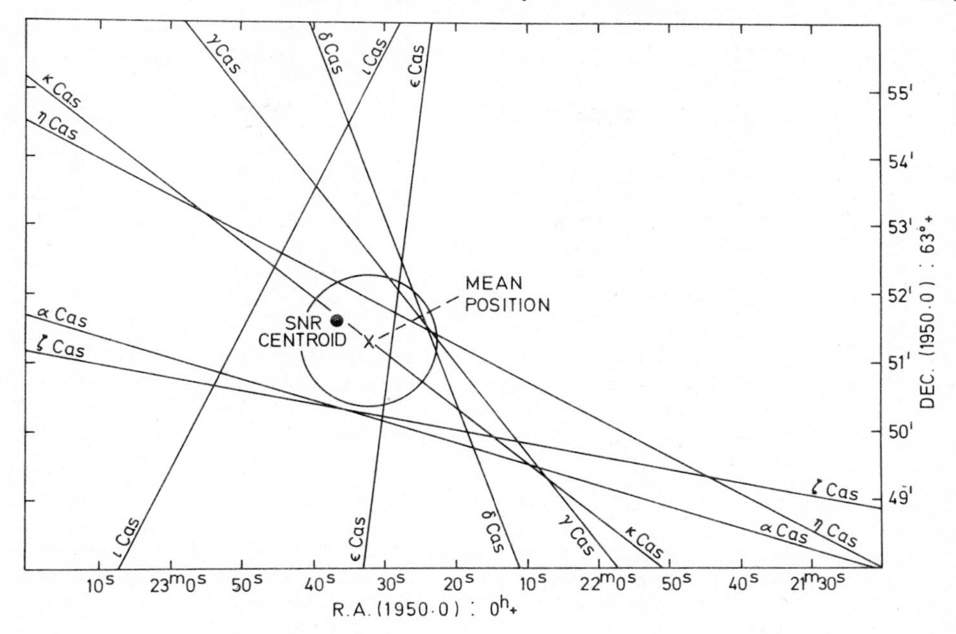

Fig. 10.5. The position of the new star, using Tycho's measurements with correction for systematic error.

Only the results of three observers deserve special mention.

Mugnoz of Valencia measured the distance of the supernova from the three stars α, β and γ Cas. His results deviated from the calculated values for the supernova remnant by 0.9', 0.8' and 11.6' respectively. Clearly he was a fairly careful observer, although not up to Tycho's standard.

The English mathematician, Thomas Digges of Cambridge, measured the angular separation of the new star from 6 stars of Cassiopeia. These were α, β, γ, δ, ε and κ Cas. He used a cross staff for this purpose. His results are summarised in Table 10.5, with the calculated values for the SNR for comparison. They are taken from chapter 9 of Tycho's Progymnasmata.

Digges' results show only a slightly greater scatter than those of Tycho, and furthermore reveal no evidence of systematic error. Tycho, comparing his own measurements with those of Digges, believed that the latter did not make adequate allowance for the effect of parallax in using his instrument. In point of fact Digges' results would give a much better location for the supernova than Tycho's. However, this conclusion is only apparent in retrospect from our knowledge of the accurate position of the centroid of the remnant. It was natural for Baade and the other early investigators of the location of the new star to concentrate on Tycho's data in view of his undoubted supremacy as an observer.

Star	Calculated	Digges	Diff.(C-D)
α Cas.	7°49.1'	7°47'	+2.1'
β Cas.	5 20.8	5 15	+5.8
γ Cas.	4 58.4	4 58	+0.4
δ Cas.	7 59.6	8 05	-5.4
ε Cas.	9 44.5	9 44	-0.5
κ Cas.	1 28.3	1 28½	-0.2

Table 10.5. Comparison between calculated and Thomas Digges' measurements of the new star from Digges' selected stars in Cassiopeia.

Thomas Digges also adopted a very simple procedure which was capable of yielding very accurate results. In order to determine whether the star had any detectable parallax, he selected two pairs of stars so that (as accurately as could be judged) the new star lay at the point of intersection of the lines (i.e great circles) joining each pair. Digges checked the alignment with a ruler 6 feet in length. He stated that the new star lay exactly at the point of intersection of the lines joining δ Cas to β Cep and ι Cep to γ Cas, and concluded that the star could not have a parallax as large as 2 arc min. Tycho, in deducing the position of the new star from Digges' data, showed that Digges had given the stars in the wrong order; the two lines must be understood to join β Cep to α Cas and ι Cep to δ Cas. This correction gave a location very near his own position.

Maestlin adopted a very similar procedure to Digges (*Prog.*, chapter 8), using a taut thread instead of a ruler. He found that the new star lay at the point of intersection of the lines joining δ Cas to ι Cep and β Cas to λ U Maj. His conclusion from this experiment was that the new star had no parallax and was situated among the fixed stars.

We have deduced the 1950.0 coordinates of the supernova by combining the observations of Digges and Maestlin (it will be noticed that both observers chose the pair of stars δ Cas and ι Cep). All three lines meet almost in a single point, having coordinates approximately $0^h23^m13^s$, +63°50.6'. However, this is nearly 4 arc min to the east of the centroid of the SNR. In view of the difficulty in determining whether the ruler or thread was absolutely straight, this seems a remarkably good result. The discrepancy is in the opposite direction to that of Tycho. To sum up, there seems more than adequate grounds for believing that the original measurements of the position of the supernova are in full accord with the modern determination of the location of the SNR centroid.

Figures 10.6 and 10.7 show high-resolution maps of the radio remnant of Tycho's supernova made with the Westerbork Synthesis Radiotelescope at frequencies of 610 and 4995 MHz respectively (Duin and Strom, 1975). At the lower frequency the resolution of the instrument is ≃1 arc min, and at the higher frequency ≃7 arc *sec*. The lower frequency map shows a well-defined ring structure characteristic of supernova remnants, although in the

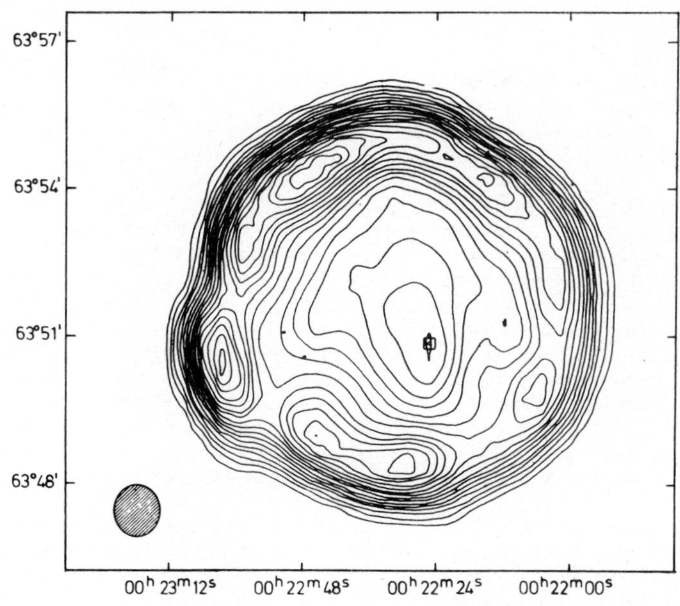

Fig. 10.6. Tycho's SNR at 610 MHz.

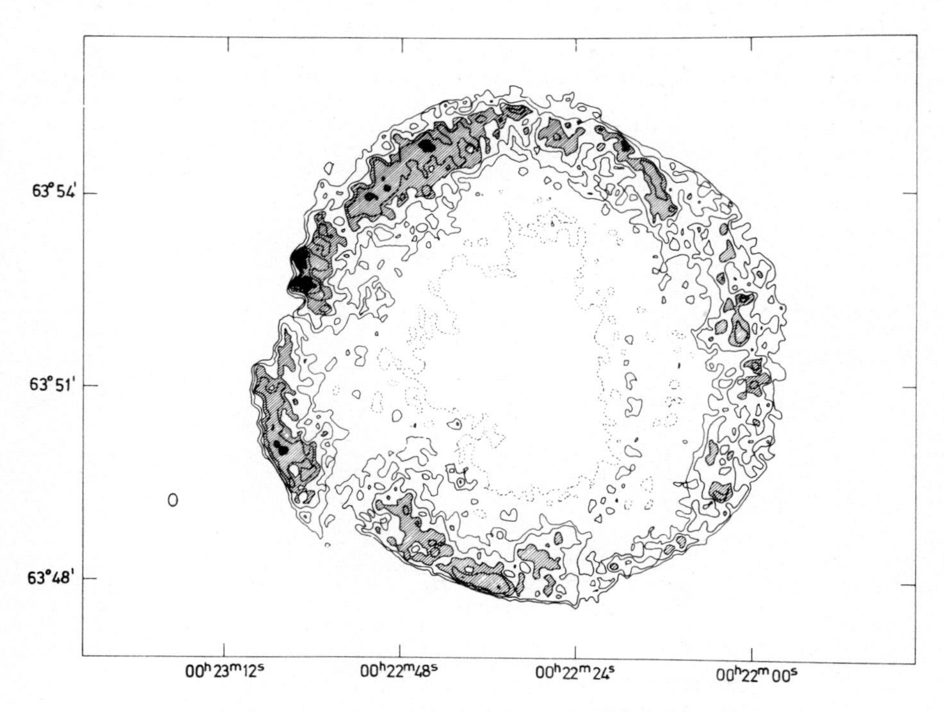

Fig. 10.7. Tycho's SNR at 4995 MHz.

high-resolution map the ring breaks up into emission "knots" - this structure is consistent with the radio emission mechanism of Gull (1973a) for young supernova remnants, where convective mixing at the interface of the ejecta and swept-up interstellar material produces local enhancements of the magnetic field and consequently radio brightness. (The central minimum of Fig. 10.7 is exaggerated because of analysis difficulties.)

The radio spectral index of the remnant has, until recently, been interpreted as steepening towards low frequencies, with $\alpha=-0.55$ at $\simeq 1000$ MHz and $\alpha=-0.9$ at frequencies <100 MHz. Roger et al (1973) suggested that the data could be fitted with two power law spectra having equal flux densities at 500 MHz, and spectral indices of -0.95 and -0.35. The new high resolution data of Duin and Strom in fact suggest that these components are spatially separated. Their spectral index varies across the source, being $\simeq-0.7$ for features associated with the outer shell, steppening to -1.0 as one proceeds radially inwards, suggesting that the high energy electrons producing the synchrotron emission are distributed in a thinner shell than the low energy ones.

Polarisation measurements at different frequencies have in the past appeared inconsistent, mainly due to the unknown effects of Faraday rotation. Duin and Strom were able to determine "zones" of rotation measure from a comparison of polarisation observations made at 1420 and 5000 MHz, and correcting for this rotation their map of intrinsic polarization is shown as Fig. 10.4. The tangential alignment of the electric field vector around the entire shell, from the outer to innermost radii, is seen to be extremely good confirming that, as in other young SNRs (e.g the remnant of AD 1006, see chapter 7), the projected magnetic field perpendicular to the line of sight is directed almost perfectly radially everywhere,

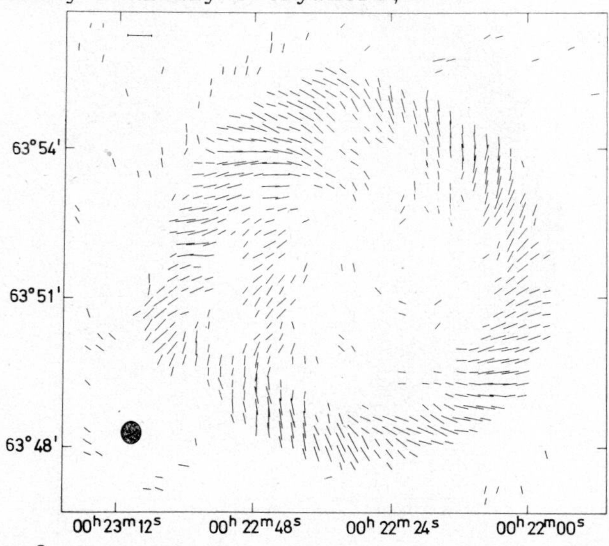

Fig. 10.8. Tycho's SNR, showing intrinsic polarization.

The distance to the remnant is stll a cause of some debate. Menon and Williams (1966) suggested a distance of ≃5 kpc from H I absorption data, giving a linear diameter of ≃11 pc and average velocity of expansion since the outburst of ≃13,400 km/sec. Comparable values had earlier been suggested by Minkowski (1964) and Woltjer (1964) from estimates of maximum brightness of the supernova. However Hughes et al (1971) suggested that the hydrogen distribution in the vicinity of the remnant was anomalous, and interpreted their H I absorption data as merely requiring a minimum distance of 1.5 kpc. More recent observations do in fact confirm earlier estimates of somewhat greater required minimum distance. Williams (1973) gives a distance ~ 5 kpc, while Goss et al (1973) interpret a weak absorption feature not apparent in some earlier observations because of observational limitations as requiring a distance $\geqslant 6$ kpc. In interpreting the dynamical evolution of supernova remnants in Chapter 12 we have cause to adopt this larger value, so that the linear diameter is ~ 13.8 pc and the average velocity of expansion since the outburst $\sim 16{,}000$ km/sec - an extremely large value, even for a Type I supernova.

Referring to Fig. 1.5, estimates of an apparent magnitude at maximum of -4 require an average interstellar absorption along the line of sight of a mere 0.2 mag./kpc assuming a typical Type I peak absolute magnitude of -19. While this value is somewhat lower than expected for this direction in the galactic plane it does not represent a serious discrepancy as to call into question the H I absorption measurement distance. Supporting evidence for such a low absorption is given by van den Bergh (1970), who estimated the reddening of the supernova from the colour descriptions in the historical records; from a comparison of the observed colours with the intrinsic colours of Type I supernovae he determined an interstellar absorption of about 1.6 magnitudes. For a distance of $\geqslant 6$ kpc, this would require a peak absolute magnitude of -19.5 or brighter. We will present later evidence suggesting that Tycho's supernova was a particularly energetic outburst, so that the peak absolute magnitude may indeed have been exceptionally large.

The optical remnant defied detection until an accurate position was determined from radio observations. Then Minkowski (1959) found extremely faint filaments seen in Hα and concentrated along the outer rim of the radio shell; these are usually interpreted as emitting sheets seen edge-on (van den Bergh et al, 1973). Only one filament is bright enough to permit spectroscopic observations. Searches for prominent emission lines such as O II and O III have been unsuccessful, and since interstellar absorption in the optical is low (see above) any such emission must therefore be intrinsically weak. Comparison of plates taken over the past two decades show that individual filaments display significant changes in brightness, and are moving radially at a rate of ≃0.2" per year. This is somewhat slower than the required average motion since the outburst of ≃0.6" per year, however it now appears that optical filaments in remnants are shocked interstellar clouds which may not directly exhibit the properties of the expanding shock wave (McKee and Cowie, 1975). In the absence of observable details near the centre of the remnant, the radial velocity of expansion cannot be determined so that it is not possible to reach any conclusions about the distance to the remnant from optical measurements.

A relatively bright star located near the centroid of the radio emission shows a late-type spectrum and is clearly not the stellar remnant. Van den Bergh et al (1973) report that attempts to obtain a spectrum of a faint bluish companion to this star have so far remained tantalizingly unsuccessful.

The remnant of Tycho's supernova is a weak X-ray source first detected by Friedman et al (1967). It is found from proportional counter data that the spectrum of the emission is well fitted by a two-component thermal model (Davison et al, 1976), with a prominent emission feature attributed to iron. Thermal emission from a shock-heated interstellar material must therefore now be accepted as the source of the X-ray emission (contrary to earlier suggestions by Shklovsky, 1974, who suggested that the X-ray source was most likely a pulsar). The emission feature requires an abundance of iron in the neighbourhood of the remnant 10 to 20 times greater than the typical interstellar value; this could be achieved by enrichment of the iron content by material ejected from the supernova having an iron abundance four times the cosmic value (Davison et al, 1976). Higher resolution spatial and spectral observation of the X-ray source are urgently required.

The historical observation reported by Tycho Brahe provide us with more primary data for the AD 1572 supernova than that combined from all previously recorded outbursts. In particular the unambiguous classification of the event as of Type I, and the positive identification of its remnant at optical, radio, and X-ray wavelengths, provide astrophysicists today with a near-ideal candidate for studying the behaviour of a "typical" (although particularly energetic) SNR. It is to be expected that observational advances will in the near future resolve any remaining uncertainties as to the distance of the remnant, the identification of the stellar remnant, the nature of the optical filamentary emission, and the structure of the X-ray source.

In AD 1596, after falling out with his patrons in the Danish Royal Court, Tycho Brahe left his native Denmark to become Imperial Mathematician of Bohemia. He was soon joined by a much younger man by the name of Johannes Kepler, who had been born just 11 months prior to the recorded outburst of the supernova which was to make Tycho famous. It must rank as one of astronomy's most fortunate coincidences that the next galactic supernova was to be sighted during Kepler's lifetime, but one of astronomy's greatest tragedies that its appearance should precede the application of the telescope to astronomical research by a mere five years.

Chapter 11

THE NEW STAR OF JOHANNES KEPLER

Many people who witnessed the AD 1572 event must have also seen this star, which occurred only 32 years later. Unfortunately, Tycho Brahe, who did so much for the earlier new star, had died in AD 1601. However, his successor as Imperial Mathematician at Prague, Johannes Kepler (AD 1571 - 1630), was to make similar detailed observations of the new star which bears his name. These enable us to prove that the star, like its predecessor, was a supernova of Type I, and conclusively identify its remnant as G4.5+6.8, an established SNR. Kepler's greatest work was undoubtedly his study of Tycho's observations of the planet Mars, which led to the formulation of his laws of planetary motion. However, his study of the supernova must also rank as one of the major contributions to science of the age.

Despite its southerly declination (-21°) and proximity to the Sun (only 58° E at discovery), the new star was detected almost simultaneously by astronomers in Europe and the Far East. It was a fortunate coincidence that the star appeared only about 3 degrees to the northwest of Mars and Jupiter (which were then in conjunction) and about 4 degrees to the east of Saturn, so that this area of sky would be under constant surveillance at this time. The supernova was discovered on the evening of October 9 (all dates are on the Gregorian calendar) independently by two Italian observers. Baade (1943) points out that "we have the emphatic statements of Fabricius and several other reliable observers that nothing extraordinary was noticed when they observed Mars and Jupiter on the evening of October 8". The star was thus still inconspicuous on October 8. By October 10, the star was sighted in China, and three days later the court astronomers of Korea (at Sŏul) first saw it. Bad weather in Northern Europe prevented observation on several successive nights. For example, although Kepler had been notified of the discovery of the supernova on October 10, he did not himself see it until a full week later. Possibly the Korean astronomers, who, as we shall learn, so carefully observed this star, were similarly troubled for the first few days.

As will be shown below, the probable date of the maximum of the supernova was around November 1, so that the star was discovered nearly 3 weeks before the maximum. Even for the systematic supernova searches maintained during the past few decades at Mt. Palomar, Asiago, etc., such a large interval is quite rare.

Conjunction of the star with the Sun occurred on December 9, but after heliacal rising, it was recovered both in Europe and in the Far East. The last definite sighting by Kepler was on October 8 in the following year, only two days after the star was last seen in China. The period of visibility was thus virtually a full year.

The position and changing brightness of the new star were measured very carefully by the European and Korean astronomers. It seems possible that similar measurements were made in China, but only brief records have survived. As is true of the AD 1572 event, there is no known reference to the star in Japanese history.

Following our usual practice, we shall begin with the Far Eastern records. Chinese observations of the new star are reported in the astronomical treatise of the Ming-shih (chapter 27) and in the Ming-shih-lu (Annals of the reign of Emperor Shên-tsung).
Ming-shih: "32nd year (of the Wan-li reign period of Emperor Shên-tsung), 9th month, (day) i-ch'ou. In the division (fên) of Wei there was a star like a crossbow pellet. Its colour was orange (i.e. reddish-yellow) and it was seen in the south-west. In the 10th month it became obscured. On (the day) hsin-yu in the 12th month it reappeared in the south-east. It was still in the division of Wei. In the 2nd month of the following year it gradually became dim. On (the day) ting-mao in the 8th month it was finally extinguished".
Ming-shih-lu: "(32nd year of the Wan-li reign period, 9th month), (day) i-ch'ou. At the beginning of the night in the south-west there arose a strange star as large as a crossbow pellet. Its body was orange (i.e. reddish-yellow) in colour. It was called a 'guest star' ".
Ming-shih-lu: "(33rd year of the Wan-li reign period, 8th month), (day) ting-mao. At night the guest star was not seen. From the 9th month of the 32nd year the guest star was visible in the division of Wei. At the first watch it regularly appeared in the south-west, and it turned with the sky. During the 10th month in the evening it sank and was not visible. In the 11th month at the 5th watch it regularly appeared in the south-east. During the 1st month of the present year its light gradually became dim and now it has finally been extinguished."

Combining these three records, the date of first appearance of the star corresponds to AD 1604 October 10. The date when the new star was "finally extinguished" corresponds to AD 1605 October 7. Heliacal setting took place in the month between November 21 and December 21 (in AD 1604). There is discord between the Ming-shih and the Ming-shih-lu over the date of heliacal rising and the month when the star "gradually became dim". The latter is of course unimportant since the interpretation of "dim" (an) is quite subjective. The Ming-shih gives the exact date of heliacal rising (equivalent to AD 1605 February 3), but this seems very late. The Ming-shih-lu date (December 22 to January 18) agrees well with the Korean date (December 22 - see below). There is an unaccountable error somewhere.

The Chinese records would alone be sufficient to prove that the star was not a comet, but only a very vague location is given, i.e. an approximate R.A., but no dec. A rough dec. might be deduced from the direction in which the star was seen, but it seems unlikely that the remnant could be located from the Chinese record alone. There is even no direct reference to the brightness of the star.

The standard of all three Chinese records is uniformly low. We find it difficult to judge whether the original observations of the imperial astronomers (at Peking) were deficient or whether the chronicles omit much valuable material in the interests of brevity.

In contrast, the regularity and detail of the Korean observations come as a major surprise. Only a very brief Korean record of SN 1572 exists, but for nearly 6 months we find almost daily reports of the new star of AD 1604. These are contained in the Sŏnjo Sillok, the annals of the reign of King Sŏnjo of Korea (AD 1567 to 1607). On reading the reports of the star in the Sŏnjo Sillok, one cannot fail to be impressed by the remarkably scientific attitude of the court astronomers. On almost every clear night the right ascension of the star and its north polar distance were measured, while its brightness was compared with that of neighbouring planets and stars, just as in Europe. We are regularly told when thick clouds blanketed the sky, ruling out observation. Frequent statements to the effect that cloud or bright moonlight, although not conclealing the star, prevented measurement of its co-ordinates, testify to the care which was exercised. Although positions are only determined to the nearest degree, after the first night the same location is reported night after night. Almost one hundred observations of the new star are recorded, and we are told that cloud either wholly or partially interfered with visibility on nearly 40 nights. This day to day diary of observations of the star is without equal anywhere else in the world.

Like most Korean literary works, the Sŏnjo Sillok is written in Classical Chinese (Hanmun). Because of this, we have preferred to give the romanisation of characters as for a Chinese work except for the name of the king (Sŏnjo), for which the Korean pronunciation is given.

The first report of the new star is as follows: "(37th year of Sŏnjo, 9th month), (day) wu-ch'en. In the first watch of the might a guest star was in the 10th degree (tu) of Wei lunar mansion and distant 110 degrees from the (north) pole. Its form was slightly smaller than Jupiter. Its colour was orange (i.e. yellowish-red) and it was scintillating. In the 5th watch there was mist".

The discovery date corresponds to AD 1604 October 13. On the following night a slightly revised position is given ("the 11th degree of Wei and distant 109 degrees from the pole), and the detail is added that the guest star was "above T'ien-chiang". Otherwise the record is virtually identical with the previous one. On all future occasions this revised position is retained, and apart from changes in the brightness of the star, essentially this same account is found on every clear night up to November 26. Between October 28 and November 5, the star was compared with Venus in brightness, and from October 28 to 31 we find the description "its ray emanations were very resplendent", suggesting that it was at maximum brilliance around this time.

After November 26 there were a few cloudy days - "there were dense clouds and the guest star was not seen". There are no mentions of the star from November 30 to December 3, and by Decembe 4 the star had already set heliacally. On this date the entry reads: "The guest star was close to the Sun. It sank in the west before dusk and could not be observed". The precise date of heliacal setting is thus unknown, but it was sometime between November 27 and December 3.

By December 27 the star was again visible. For this day we read, "At daybreak the guest star was seen in the east above the stars of *T'ien-chiang*. It was within the 11th degree of *Wei* and distant 109 degrees from the pole. Its form was larger than Antares. Its colour was orange and it was scintillating". Apart from slight textual variations and the diminishing brightness of the star, precisely the same account occurs on every clear night except when the Moon was bright until April 5. After this date, references to the star are few and far between, presumably because the original observations had been lost. The single record on April 23 follows the usual pattern, while on May 2 we read, "at the 4th watch the guest star was faintly visible through rifts in the clouds". Subsequently, there are only three indirect references to the star, two commenting on its astrological significance, and the last, dated September 15, stating that "there were dense clouds and the guest star was not seen". It thus seems probable that the star continued to be observed for a similar lengt of time to that in China and Europe (i.e. for virtually a full year).

The Korean measurements of position are interesting in that they prove that the position of the star remained fixed for nearly 6 months. However, their accuracy is far inferior (by two orders of magnitude) to the European measurements. Nevertheless, we have thought it of interest to compare the Korean location with that derived from the European measurements (see below). From chapter 2, the determinant star of *Wei* is μ^1Sco. The R.A. of the new star thus reduces to $17^h\ 06.6^m \pm 2^m$. This compares with the best European result of $17^h\ 07.1^m$. The Korean position is thus tolerabl accurate. The polar distance (109 *tu*) corresponds to a declination of -17.4^o (European result -21.1^o) so that the error is thus rather large (some 4 degrees). If this distance was measured from Polaris, which would presumably be the pole star of the time, the error would be slightly larger, so that this is not an explanation.

A remarkable observation made by the Korean astronomers on January 20 in AD 1605 allows us to fix the position of the star with a fair degree of accuracy from the oriental data alone. The appropriate entry in the *Sŏnjo Sillok* reads : "(37th year of Sŏnjo, 12th month), (day) *ting-wei* ...... Venus invaded the guest star (*chin-hsing-fan-k'o-hsing*)". As discussed in Chapter 9, the expression *fan* ("to invade", "trespass against", etc.) was an astrological term implying an approach of two celestial bodies to within about a degree of one another. Calculation based on the

European position shows that before dawn on January 21 (astronomical date January 20 - see Chapter 9), Venus and the new star would be only 0.51 degrees apart. In the absence of the European measurements this single observation would probably have been sufficiently accurate to allow the remnant of the new star to be identified with confidence. Fortunately, thanks to the efforts of Kepler and Fabricius, this is no more than an academic exercise.

By far the most important Korean observations are the brightness comparisons. Before heliacal setting the brightness was compared with the planets: Jupiter (<u>sui-hsing</u> - "Year Star"), Venus (<u>chin-hsing</u> - "Metal Star") and Mars (<u>hui-hsing</u> - "Fire Star"). After heliacal rising, only two reference stars were chosen : (α Sco. (<u>Hsin-ta-hsing</u> - the "Great Star of <u>Hsin</u>") and τ Sco. (<u>Hsin-tung-hsing</u> - the "Eastern Star of <u>Hsin</u>"). The lunar mansion <u>Hsin</u> consists only of three stars, with τ Sco at the east, α Sco (Antares) in the middle and σ Sco at the west so that the Korean names are unambiguous. The new star was described as "larger than" (<u>ta-yu</u>) Antares between December 26 and January 20, "as large as" (<u>ta-ji</u>) Antares between January 21 and February 15, "slightly smaller than" (<u>ch'a-hsiao-yu</u>) Antares between February 19 and 21, and finally "smaller than" (<u>hsiao-yu</u>) Antares between February 23 and April 5. Eventually, on April 23 the star was described as "as large as" (<u>ta-ju</u>) τ Sco.

We have corrected two minor presumed errors relating to January 30 and February 18. On both these dates the star was described as "larger than" (<u>ta-yu</u>) Antares, which does not conform with the general trend. Possibly these are textual errors. Alternatively, local atmospheric conditions could have made the new star seem rather brighter than Antares on each occasion. It seems best to regard both reports as spurious.

It so happens that Antares was an unfortunate choice for it is a semi-regular variable. The approximate period is 5 years and the extreme range of magnitude, reduced to the Harvard system, +1.1 to 2.0 (Kukarkin et al, 1969). In studying the light curve of the new star we have assumed this same range of brightness.

The various brightness comparisons are summarised in Table 11.1.

In Table 11.1 dates do not always run continuously, either because of gaps in the observations (due usually to cloud) or in the records themselves (presumably due to loss of the original reports). Although the new star was compared with Venus during the period October 28 to November 5, at that time the planet was a morning star (on the opposite side of the Sun). Direct comparison, unless in daylight- and there is no reference to the visibility of the new star during the daytime - was thus impossible, and the only sound inference seems to be that the star was considerably brighter than Jupiter. This is how we shall interpret the observations.

Mars was the reference object in the last few days before heliacal setting (November 20 - 26). However, the planet was then about 50 degrees east of the Sun while the star was rapidly approaching

DATE OR DATE RANGE		BRIGHTNESS LEVEL
AD 1604	Oct. 13/14	Smaller than Jupiter
	Oct. 15/16	Slightly smaller than Jupiter
	Oct. 17/18	As large as Jupiter
	Oct. 19-25	Larger than Jupiter
	Oct. 28-31	As large as Venus, rays prominent
	Nov. 1- 5	As large as Venus
	Nov. 7-13	As large as Jupiter
	Nov. 14	Slightly smaller than Jupiter
	Nov. 16-19	Smaller than Jupiter
	Nov. 20-23	As large as Mars
	Nov. 24-26	Smaller than Mars
	Dec. 26-Jan. 20	Larger than Antares
AD 1605	Jan. 21-Feb. 15	As large as Antares
	Feb. 19-21	Slightly smaller than Antares
	Feb. 23-Apr. 5	Smaller than Antares
	Apr. 23	As large as τ Sco

Table 11.1. Comparisons of the brightness of the new star with planets and fixed stars by the Korean astronomers.

conjunction (elongation only 13 degrees east on November 26). The relative visibility of the new star would thus be very much impaired by atmospheric extinction (on account of its low altitude) and the twilight glow, giving the impression that it was much fainter than it actually was. When the comparisons with Jupiter were made, planet and star were fairly close together so that both would be viewed under much the same conditions.

Before leaving the oriental records, it seems appropriate to make some reference to the "night watches" (*kêng*) which were in use throughout the Far East. Generally speaking, the night between dusk and dawn was subdivided into five watches, presumably of equal length. From Needham, Wang-ling and Price (1960), the durations of dusk and dawn were each assumed to be 2½ *k'o* (36 minutes) - similar to our "lighting up time". Normally the remainder of the interval between sunset and sunrise would be taken up by the watches. The average length of a watch would thus be about 2 hours 10 minutes, but the actual length would obviously vary with the season. In this respect, the night watches were similar to "seasonal hours" which were used throughout the ancient and medieval West.

Coming now to the European observations of the new star, Kepler made a detailed study of the object in his *De Stella Nova in Pede Serpentarii*. However, whereas Tycho Brahe collected practically every significant observation of the supernova of AD 1572 made by his contemporaries, the literature relating to the star of AD 1604 is very scattered. The best reference work is undoubtedly the paper by Baade (1943). Baade searched carefully through early 17th century European astronomical writings for observations. As a result of this search, he felt that he had probably isolated all important documents with the exception of the reports of David Fabricius, which were not accessible to him. However, Baade was able to obtain a few of the observations made by Fabricius from the latter's correspondence with Kepler so that this omission was probably not too serious.

It seems that the new star first appeared on the evening of October 9. On that day it was reported independently by two Italian observers - a physician in Cosenza whose name is unknown and I. Altobelli in Verona. The physician reported his observations of the star to the Jesuit astronomer Clavius in Rome. As already noted, we have the testimony of Fabricius and others, who observed Mars and Jupiter on the evening of October 8, that the star was not then evident. By October 10, the new star was seen by two observers in Padua, and also in Prague by J. Brunowski, who notified Kepler of his discovery. Baade collected the various estimates together in a table. This is reproduced as Table 11.2, but we have excluded the names of the observers and the few references to the colour of the star. On October 9 the physician reported the colour of the star as "like Mars", and on the same day Altobelli made the analogy "like half of a ripe orange". Again Capra-Marius of Padua on October 10 made a comparison with Mars, but by October 15, both the physician and Fabricius described the star as "like Jupiter", Fabricius adding that it was "white, not red". These latter colour estimates are at variance with those made in Korea, which were uniformly "orange" from discovery until AD 1605 April 5. On April 23 the colour was said by the Korean astronomers to be "yellow", but possibly by this time the star would be so faint that colour judgement would be unreliable.

Date		Brightness estimate
AD 1604	Oct. 8	Not seen
	Oct. 9	As bright as Mars
	"	As bright as Jupiter
	Oct. 10	Somewhat brighter than Mars
	"	Very similar in brightness to Mars
	Oct. 11	Still brighter than on Oct. 10 when somewhat brighter than Mars.
	"	Twice as bright as Jupiter
	Oct. 12	Almost as bright as Jupiter
	Oct. 15	As bright as Jupiter or somewhat more
	"	Much brighter than Jupiter; no further increase after this day
	"	As bright as Jupiter or a little more; no further increase afterward
	"	A little brighter than Jupiter
	"	As bright as or brighter than Jupiter
	"	Brighter than Jupiter and equal to Venus
	Oct. 17	Much brighter than Jupiter (almost twice as bright)
AD 1605	Jan. 3	Brighter than α Sco, much fainter than α Boo
	Jan. 13	Brighter than α Boo and Saturn
	Jan. 14	About as bright as Mars (in Oct. 1604)
	Jan. 21	About as bright as α Sco, a little brighter than Saturn
	End of Jan.	As bright as α Vir
	Mar. 20	Not much brighter than ζ and η Oph
	Mar. 27	Not much brighter than ζ and η Oph
	Mar. 28	Not much brighter than η Oph
	Apr. 12	As bright as η Oph
	Apr. 21	As bright as η Oph
	Aug. 12-14	As bright as ζ Oph
	Aug. 29	About as bright as ζ Oph
	Sep. 13	Fainter than ζ Oph
	Oct. 8	Difficult to see; fainter or equal to ζ Oph

Table 11.2. European estimates of the brightness of the new star.

Most of the later observations in Table 11.2 (after the beginning of AD 1605) were made by Kepler. There do not appear to be any useful European estimates of the brightness of the star from October 17 until January 3. Kepler (De Stella Nova, chapter 1) judged that the star "was seen with almost the same magnitude during the whole of the month of October". Before conjunction with the Sun he last saw the star on the evening of November 16.

He was frustrated by cloud on the next few days, and by November 22, although, as he tells us, the position of the star was indicated by the presence of the Moon (then a young crescent), he could not see it. Kepler concluded that from Prague (50° N) the new star set heliacally between November 16 and 22. In Korea, the date of heliacal setting was some time between November 27 and December 3, but the oriental astronomers had the benefit of a much more southerly location ($37\frac{1}{2}^{o}$ N).

It was not until January 3 (AD 1605) that Kepler rediscovered the star (after conjunction with the Sun). This was a full week after the Korean astronomers sighted it, but apart from his inferior situation, Kepler had been clouded out for some time. He noted that "it was much lessened from its original magnitude". After this we only get occasional bulletins regarding the brightness of the new star, usually at intervals of about a month, but there is a major gap between April 11 and August 12 during which apparently he made no observations. Kepler does not appear to have observed the star anything like so regularly as Tycho Brahe some years before.

The last positive sighting was on October 8 in AD 1605. Kepler wrote: "Now exactly a year after its first apparition in a very clear sky, its appearance could be noted only with difficulty ... From this time anew it set heliacally".

At the beginning of AD 1606 the inclement weather gave Kepler few opportunities to search for the star. He looked carefully for it on the morning of January 26, and again on February 4 but could not detect it. His comment on February 6 is interesting: "I left the observatory, not sure whether I had seen any trace of the new star. Therefore it seems to have become too small to be seen even on this clear morning, if it has survived". After a further search in March, having allowed time for the elongation of the star from the Sun to increase, he realised that it had ceased to be visible. His final conclusion was that "It is therefore uncertain on which day between October in the year 1605 and February in the year 1606 it had vanished".

In both the European and Korean brightness estimates we find expressions equivalent to some of the following: "much brighter than"; "brighter than"; "slightly brighter than"; "as bright as"; "slightly fainter than"; "fainter than". In making his magnitude reductions, Baade adopted a photometric step of 0.25 mag. between each of the above intervals. Thus we have the following conversions:

Much brighter than	-0.75 mag.
Brighter than	-0.5
Slightly brighter than	-0.25
As bright as	0
Slightly fainter than	+0.25
Fainter than	+0.5

Baade justified his choice of magnitude interval in two ways. Modern variable star observers adopt a photometric step of 0.1

mag., but such a high accuracy would be unlikely to be achieved by relatively untrained observers. A better estimate is provided by the smallest magnitude interval used in European star catalogues around AD 1600. This is one-third of a magnitude, which Baade considered more appropriate. In practice he reduced this to a quarter of a magnitude since the supernova was mostly compared with nearby stars. For the time near maximum, Baade made an attempt to allow for the rapid increase in the width of the photometric step at higher brightness; he multiplies all magnitude intervals by 1.5. To provide an independent check by an experienced variable star observer, we asked Dr. Patrick Moore of Selsey Observatory, England, to comment on the Korean data. He suggested magnitude conversions which were in excellent agreement with Baade' choice. We have accepted Baade's figures without further question.

Two factors which Baade somewhat surprisingly ignored in his magnitude reductions are the effect of moonlight and differential atmospheric extinction. A close approach of the Moon (particularly when full) to either the supernova (which was very near the eclipti or the comparison object could appreciably alter the relative visibility. However, with the aid of the tables of Tuckerman (1964), we have verified that on the dates of all the available European observations the Moon did not come within 30 deg. of the supernova or reference planet/star so that its effect can be ignored. The Korean observations were, of course, made almost daily. However, because of this, the only important records are those for which changes in brightness are noted in the middle of a period when the star and compared object were judged of equal brightness. On these critical dates, interference by moonlight was again negligible.

Differential atmospheric absorption between the supernova and comparison object on a number of occasions proves to be significant. This largely results from the low meridian altitude of the star in both Europe and Korea. We have examined each individual observation to allow for differential atmospheric absorption as far as possible. In this we have made use of the tables of Bemporad (1904), as published by Schoenberg (1929). Bemporad's tabular values have been amended slightly by us to correspond with atmospheric transmission coefficient of 0.80 quoted by Allen (1973).

Visual magnitudes of the various comparison stars are on the Harvard system (cf. Schlesinger, 1940). Obviously, as the object is to produce a light curve, the precise magnitude system is relatively unimportant. We have checked on the variability of each star using the catalogue of Kukarkin et al (1969). Apart from α Sco, already discussed, all reference stars can be regarded as of constant brightness. Planetary magnitudes -also on the Harvard System - were computed by Baade and we have accepted these without alteration. The calculated visual magnitudes of the new sta are listed in Tables 11.3 and 11.4, treating Korean and European data separately. In each case the comparisons with Venus are understood to mean "much brighter than Jupiter". Observations not includ- ed in these tables are the Korean comparisons with Mars and the

European ones with Jupiter on October 9 and 10, α Boo on January 3 and Mars on January 14. The Korean observations have already been discussed. Baade justifiably rejected the Jupiter comparisons on October 9 and 10 since these differ from those of the other observers by more than 2 mag. Again, the α Boo comparison differs by more than 2 mag. from neighbouring estimates (allowing for atmospheric extinction). Finally the reference to Mars on January 14 must be rejected since it refers to the brightness of the planet as it was 3 months previously (in October of AD 1604).

In the tables, days are counted after AD 1604 October 8 and magnitudes are estimated to the nearest 0.05.

Day No.	Corrected mag.
6	- 1.1
8	- 1.45
10	- 1.8
11	- 2.55
20	- 2.95
28	- 2.95
33	- 1.95
37	- 1.7
39	- 1.35
104	+ 0.8 (mean)
119	+ 1.55(mean)
134	+ 1.95(mean)
138	+ 2.3 (mean)
198	+ 2.9

Table 11.3. Reduction of Korean brightness estimates to magnitudes.

Day No.	Corrected mag.
0	+ 3 or fainter
1	+ 0.9
2	+ 0.5
3	- 0.7
4	- 1.5
7	- 2.2
9	- 2.6
87	+ 1.05 (mean)
97	0.0
105	+ 1.55 (mean); + 0.5
114	+ 1.2
164	+ 2.25
171	+ 2.25
172	+ 2.25
187	+ 2.40
196	+ 2.45
310	+ 4.45
326	+ 4.45
341	+ 4.95
366	+ 4.7

Table 11.4. Reduction of European brightness estimates to magnitudes.

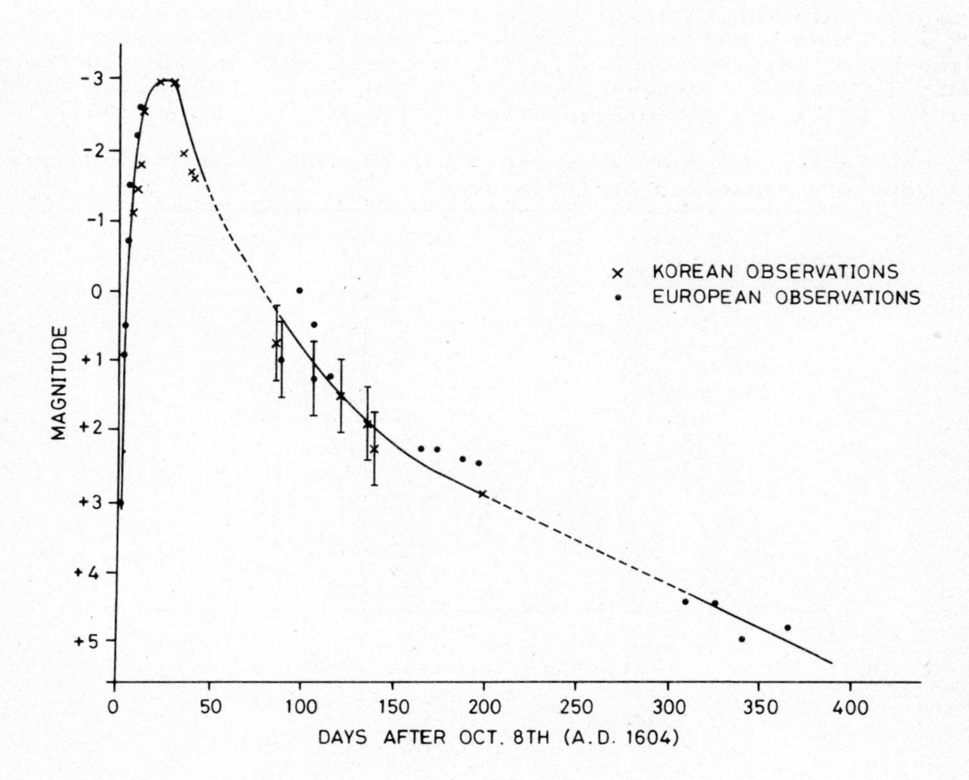

Fig. 11.1 The light curve of the supernova of AD 1604 constructed from European (shaded circles) and Korean (crosses) observations.

The various estimates are shown diagrammatically in Fig. 11.1. Korean observations are denoted by crosses, European observations by shaded circles. In the case of the Antares comparisons, the range of magnitudes corresponding to the probable extremes in brightness of the variable are indicated. The light curve obtained from such a varied assortment of crude estimates is remarkably well defined. This is clearly characteristic of a Type I supernova, as for the new star of AD 1572.

A fair estimate of the apparent magnitude of the supernova at maximum is around -3.0. The date of maximum was probably some 20 days after October 8, and thus about October 28 (in 1604).

From the form of the light curve, the supernova nature of the new star is well established. There remains the question of the remnant. The angular distance of the supernova from the stars α Aql (Altair), α Oph, η Oph, ξ Oph, α Sco (Antares) and σ Sgr was carefully measured by Kepler. Fabricius made similar measurements for all of the above stars except σ Sgr. From this data the position of the new star was deduced by Schlier (1934) and Böhme (1937). The former used only the measures of Fabricius, having demonstrated, somewhat surprisingly, that these were far superior to Kepler's data. His result, reduced to the epoch 1950.0 is: R.A. $17^h27^m39.1^s \pm 3.1^s$; dec. $-21°26'39'' \pm 42''$. Böhme, after he corrected some "obvious misprints" in Kepler's results, solved from both sets of data. He obtained a position almost identical with that derived by Schlier: $17^h27^m38.5^s \pm 2.0^s$; $-21°26'38'' \pm 26''$ at epoch 1950.0.

Using the above positions, Baade (1943) made a search for an optical remnant. He used red light because the supernova appeared in a heavily obscured part of the Milky Way. He was immediately successful; the very first plate revealed a faint patch of nebulosity whose centre lay only 2.1^s W and 1" N of the position of the new star. The positional agreement is thus far superior to that in the case of Tycho's supernova.

Listed as number 358 in the third Cambridge catalogue, the remnant of Kepler's supernova is a strong non-thermal radio source of $\sim$3 arc min diameter. Its detailed structure remained essentially uncertain until recently, since the southerly declination puts it beyond the range of objects that can be properly mapped by high-resolution aperture synthesis radiotelescopes in the Northern Hemisphere; (only now are Southern Hemisphere synthesis radiotelescopes being built with sufficient resolution to study sources as small as Kepler's SNR). The beam-shape of an attempted high-resolution synthesis map by Hermann and Dickel (1973) was so distorted as to make the map almost impossible to interpret. A lunar occultation measurement by Hazard and Sutton (1971) was reported to be consistent with the source having a broken shell of diameter 3 arc min. The true structure of the source was eventually revealed from 5000 MHz observations made using the Cambridge one-mile radiotelescope and the Owen's Valley interferometer and analysed by Gull (1975). Although each set of observations independently suffered from limited coverage of the aperture plane, the data could be very

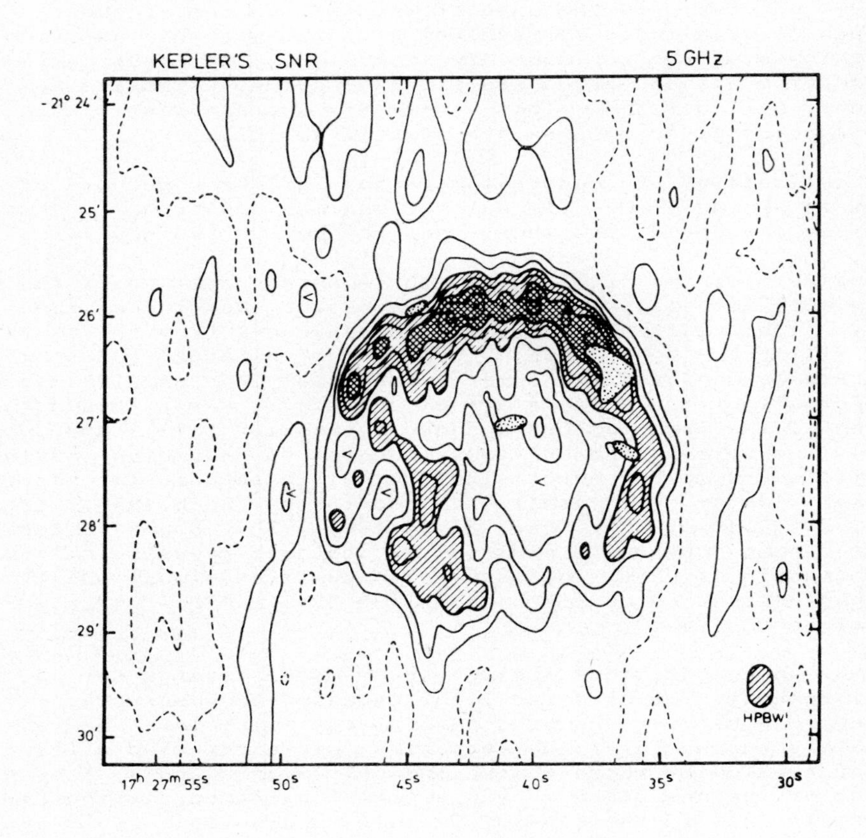

Fig. 11.2 A 5000 MHz map of the radio remnant of Kepler's supernova, constructed by Gull (1975) from observations made with the Cambridge one-mile radiotelescope and the Owen's Valley interferometer.

carefully combined and analysed to produce the map depicted as Fig. 11.2.

Gull's map of the remnant shows the shell structure characteristic of SNRs (and hinted at by the earlier lunar occultation observations). Although almost circular, the shell shows peak brightness along the northern edge, with a large gap to the south and irregular indentations to the east (possibly due to encounters with irregularities in the interstellar medium). The radio properties again appear to be consistent with the preferred generation mechanism of Gull (1973a) for young SNRs.

The centroid of the radio emission, R.A. (1950) $17^h\ 27^m\ 41^s$, Dec. (1950) $-21^\circ\ 27'$, is in spectacular agreement (differing by less than 30 sec of arc) with the positional determination of Fabricius given above.

The radio spectrum of the remnant is well determined. Milne (1969) has summarised data from observations in the range 86 to 5000 MHz which indicate a non-thermal spectral index of $\alpha = -0.58$. However there is no evidence for strong polarisation in Kepler's SNR (Milne and Dickel, 1974).

There are no observations which enable an unambiguous estimate of distance to the source. It is too close to the Galactic centre for H I absorption measurements to produce a meaningful determination, and the optical remnant (discussed below) is too diffuse to obtain transverse motions with sufficient precision to obtain the distance by combining them with radial velocity estimates (Minkowski, 1968). On the assumption of typical Type I behaviour, a distance in excess of 10 kpc is implied in the absence of any interstellar absorption, (see Fig. 1.5). While at the relatively high galactic latitude of the source absorption should not be excessive, some allowance for it must be made. According to Minkowski (1964), on his interpretation of the historical colour descriptions the reddening of Kepler's supernova was between 0.0 and 0.4 magnitudes greater than for Tycho's supernova, although we have already commented on discrepancies in the colour descriptions so that the method could not be considered reliable. Nevertheless, assuming a total interstellar absorption in the range 1 to 2 magnitudes would place the supernova at a distance of between 6.5 to 10 kpc. It is impossible to be more definite, although deceleration of the remnant suggests that it is expanding into a region of interstellar density characteristic of the galactic plane despite its large z-distance (~ 1 kpc). It would therefore seem an attractive hypothesis to place the supernova in the nuclear-bulge of the galaxy. In subsequent analysis we have accordingly adopted a distance of 10 kpc (with considerable uncertainty).

A search for the optical remnant of Kepler's supernova by Baade (1943) revealed a number of bright 'knots' (depicted in Fig. 11.2), with a few faint filaments. Spectral investigations by Minkowski (1943) showed emission lines of O I , O III , N II , S II , and H . Intercomparison of plates of the nebula taken with the 200 " Mt. Wilson telescope over the past 20 years shows long-term variations in the brightness of various features, and a radial movement of the brightest knot at $\sim$0.03" per year (which at the assumed distance of 10 kpc corresponds to a velocity of $\sim$1400 km sec^{-1}). This velocity is considerably less than the required average velocity of expansion since the outburst. A similar discrepancy was previously noted for Tycho's SNR, and the explanation for this suggested in Chapter 10 is presumably applicable here also. Atempts to identify the stellar remnant of the outburst have so far been unsuccessful.

Kepler's SNR is the only 'young' remnant on the near-side of the galaxy which has not yet been detected in X-rays. However this is hardly surprising in view of its extreme distance placing it beyond the sensitivity range of the present generation of X-ray instruments. Additionally, there are several strong X-ray sources in this region of sky which would produce confusion problems in any attempted observation.

No supernova has been observed in our galaxy since Kepler's time. The nearby (distance $\sim$3 kpc) strong non-thermal radio source Cas A (G111.7-2.1) has the morphological and spectral characteristics of an SNR, and is believed to be the remnant of a supernova which occurred a mere 2 to 3 centuries ago. However there is no historical record which could possibly correspond to this event. A possible explanation for its non-detection could be that it occurred in a region of anomalously high absorption ($\sim$2.5 magnitudes per kpc) so that despite its proximity to Earth it did not reach the brightness required to be readily discovered. Even allowing for such an effect, the fact that it was not seen in Europe or the Far East despite being circumpolar remains something of a mystery.

The next galactic supernova is eagerly awaited by the world's astronomical fraternity, and the means of observing such an event are now very powerful indeed. There can be no doubt that a supernova outburst would precipitate urgent and concentrated astronomical observation on a scale never before undertaken. Professor Sidney van den Bergh of the David Dunlap Observatory, University of Toronto has already circulated plans to astronomers throughout the world of observational requirements following a galactic supernova. The immediate impact on modern astronomy would undoubtedly be considerable - the long-term impact on civilization on the planet Earth of a particularly nearby supernova could be extremely dramatic, as we will speculate in Chapter 12. In the mean time the world's astronomers and astrophysicists wait, hoping that during their life-time they may be privileged to witness one of the Universe's greatest spectacles.

Chapter 12

SOME THOUGHTS ON THE EVOLUTION OF SUPERNOVA REMNANTS

In the preceding chapters we have presented eight proposed associations of historically recorded new stars with SNRs; these associations, varying from certain to possible, are summarized in Table 12.1

Table 12.1

Supernova	Radio Remnant	Remarks
AD 185	G315.4-2.3	Probable
AD 386 (?)	G11.2-0.3	Possible
AD 393	G348.5+0.1 or G348.7+0.3	Possible
AD 1006	G327.6+14.5	Certain
AD 1054	G184.6-5.8	Certain
AD 1181	G130.7+3.1	Probable
AD 1572	G120.1+1.4	Certain
AD 1604	G4.5+6.8	Certain

As mentioned in Chapter 1, most of the 120 galactic radio sources thought to be the remnants of supernovae are probably so old that no records of the outburst can be expected to exist. Thus the few identifications of historical supernovae with remnants must provide valuable observational evidence to test current theories on the evolution of SNRs.

The most straightforward method of investigating the dynamical evolution of SNRs is to use the measured diameters and current expansion velocities of those whose ages are reliably known. This method has been of limited success in the past because of the paucity of historical supernova-SNR associations, and the currently accepted evolutionary picture has been derived chiefly from theoretical work and indirect measurements. It now seems worthwhile assessing what new light the six certain or probable associations given in Table 12.1 cast on the evolution of SNRs.

The picture of the evolution of SNRs which emerges from theoretical investigations was discussed briefly in Chapter 4. After a few hundred years of near-free expansion, the evolution resembles that of an adiabatic blast wave created by releasing energy at a point in a homogeneous gas. The diameter D(pc) of the shock wave preceding the expanding shell of swept-up interstellar material is then described by the Sedov (1959) similarity solution:

$$D = 4.3 \times 10^{-11} \left(\frac{E_o}{n} \right)^{\frac{1}{5}} t^{\frac{2}{5}} \tag{1}$$

where t(yr) is the time elapsed since the explosion, E_o (erg) is the energy released in the outburst, and n (cm^{-3}) is the number density of H atoms in the interstellar medium.

The shortage of reliable age calibrators has made the direct investigation of the D - t relationship (1) difficult. For this reason the radio surface brightness Σ, which easily lends itself to observational investigation, has usually been used in evolutionary studies. It is then postulated that the variation of this parameter with linear diameter and time may be approximated by the relationships

$$\Sigma = AD^{\beta} \tag{2}$$

$$\Sigma = Bt^{\gamma} \tag{3}$$

These two expressions may be combined to give the relationship for variation of linear diameter with time

$$D = Ct^{\delta} \tag{4}$$

Because of the difficulty in evaluating the Σ - t and D - t relationships, cumulative distributions are introduced:

$$N(\Sigma) = P\Sigma^{3} \tag{5}$$

where $N(\Sigma)$ is the number of SNRs with surface brightness greater than Σ, and

$$N(D) = QD^{\eta} \tag{6}$$

where N(D) is the number of SNRs with diameter less than D. If a constant rate of supernova outbursts is assumed, then

$$N(\Sigma) = P\Sigma^{3} = \frac{t}{\tau} \quad \text{and} \quad N(D) = QD^{\eta} = \frac{t}{\tau}$$

where τ is the characteristic interval between supernova outbursts. Thus the functional dependence of Σ or D on t may be found from the cumulative distributions and only the constant τ requires calibration using SNRs of known age and measured Σ or D. The above relationships have been used in a number of SNR studies, and are shown above in the form adopted by Mills (1974).

The Σ - D relationship derived at a frequency of 408 MHz by Clark and Caswell (1976) is shown as Fig. 12.1.

$$\Sigma_{408} = 10^{-15}\ D_{pc}^{-3}$$

$$\text{for } \Sigma > 3 \times 10^{-20}\ \text{Wm}^{-2}\ \text{Hz}^{-1}\ \text{Sr}^{-1}$$

and

$$\Sigma_{408} = 3.6 \times 10^{-5}\ D_{pc}^{-10}$$

$$\text{for } \Sigma < 3 \times 10^{-20} \text{ Wm}^{-2} \text{ Hz}^{-1} \text{ Sr}^{-1}$$

It should be emphasized that the Σ - D approach is at present an empirical one with little theoretical foundation, but the quality of present observational data and their statistical implications provide valuable constraints against which future theoretical models may be tested.

The statistical investigation by Clark and Caswell (1976) confirmed that the functional dependence of D on t suggested by the Sedov equation (1) is observed in practice for the majority of SNRs. Additional data, and in particular the frequency of occurrence of supernovae $f = 1/\tau$, are required to determine the constant of proportionality; one method is to estimate the rate of occurrence of supernovae using all those of known ages, another uses the individual sizes and ages of historically recorded supernovae, and a rather different approach utilizes the X-ray data for older supernova remnants.

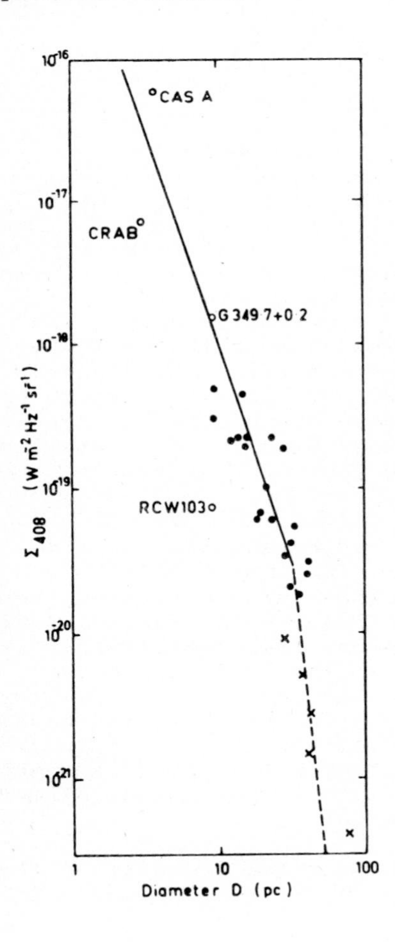

Fig. 12.1. The surface brightness vs. linear diameter curve of Clark and Caswell (1976). The solid line was determined by the calibrators indicated by filled circles. Those shown with open circles are above the surface brightness range of interest or show anomalous properties. At low surface brightness values the five calibrators indicated by crosses are all of low accuracy, but nevertheless all suggest a steepening of the curve.

From consideration of all three approaches, Clark and Caswell (1976) concluded that the best estimate of E_o/n was 5×10^{51} erg cm^3, a somewhat greater value than assumed in most previous studies. Clark and Culhane (1976) found a slightly smaller value from a detailed consideration of the third method alone. However, both investigations are consistent with a relationship

$$D \quad 0.9t^{2/5} \tag{7}$$

previously quoted in Chapter 4.

The complete set of evolutionary relationships in the form of equations (2) to (6) derived by Clark and Caswell (1976) for $\Sigma > 3 \times 10^{-20}$ Wm^{-2} Hz^{-1} Sr^{-1} is

$$\Sigma = 10^{-15} D^{-3}$$

$$D = 0.9\ t^{2/5}$$

$$N = 8 \times 10^{-3} D^{5/2}$$

$$N = 2.5 \times 10^{-15} \Sigma^{-5/6}$$

$$\Sigma = 1.25 \times 10^{-15} t^{-6/5}$$

Turning now to the individual historical supernovae:
The remnants of Tycho's and Kepler's supernovae have current linear diameters of 13.8 and 9.3 pc (see Table 4.2) and ages of 403 and 371 years respectively; inserted in equation (7), these yield values of E_o/n of 21×10^{51} erg cm^3 for Tycho's supernova and 3.4×10^{51} erg cm^3 for Kepler's supernova. Both radio remnants showing characteristic peripheral brightening, are of slightly below average Σ relative to the mean Σ - D relationship. The fact that E_o/n appears to be slightly above average for Tycho's supernova could result from an over estimate of its distance; however reducing the distance estimate could cause the value of Σ to lie somewhat further below the mean Σ - D curve. Since the distance seems well established (Goss et al, 1973), at least as a lower limit, it seems more likely that E_o/n is simply exceptionally large.

It has previously been noted that the radio remnants for the supernovae of AD 1181 (3C58) and AD 1054 (the Crab Nebula) are atypical, in that both show central brightening and flatter than average non-thermal spectra with significant polarisation confirming a synchrotron origin for the emission. Because of other enigmatic differences, the discussion of the Crab Nebula is deferred until later. In the case of 3C58, despite not displaying the peripheral brightening of SNRs, adoption of the outer boundary of radio emission as the shock front diameter leads to $E_o/n = 2.8\times10^5$ erg cm^3, in surprisingly good agreement with most other remnants. In addition, the values of surface brightness and linear diameter fit well the proposed Σ - D evolutionary track.

Applying the Σ - D relationship to the proposed remnants of the supernovae of AD 1006 and AD 185 gives distances of 4 and 3.2 kpc respectively with ages > 7000 years based on the measurement of radio surface brightness. Since we believe the historical identifications to be correct, then clearly these remnants are atypical in either their D-t evolution or their Σ - D evolution. Since we have argued that the AD 1006 supernova was only about 1 kpc from the Earth, and the AD 185 supernova no more than 2 kpc from the Earth this would indicate that it is the Σ - D evolution which is atypical; i.e. the sources are sub-luminous in their radio surface brightness. The value for E_o/n for AD 1006 is then 2.5×10^{51} erg cm^3; for AD 185, if 2 kpc is taken as an upper limit to the distance, then $E_o/n \sim 10^{52}$ erg cm^3.

The above remnant age and diameter data are plotted in Figure 12.2, together with the relationship $D = 0.9t^{2/5}$, consistent with the statistical analysis of Clark and Caswell (1976), and the X-ray data discussed in Clark and Culhane (1976). The available data confirm that, with the exception of the Crab Nebula, the remnants of the historical supernovae follow the evolutionary track suggested by the Sedov solution.

Returning to the Crab Nebula, this has long been recognised as a remarkable object, and many of its properties appear to be unique. Its current expansion velocity is ~ 2000 km sec : there is no

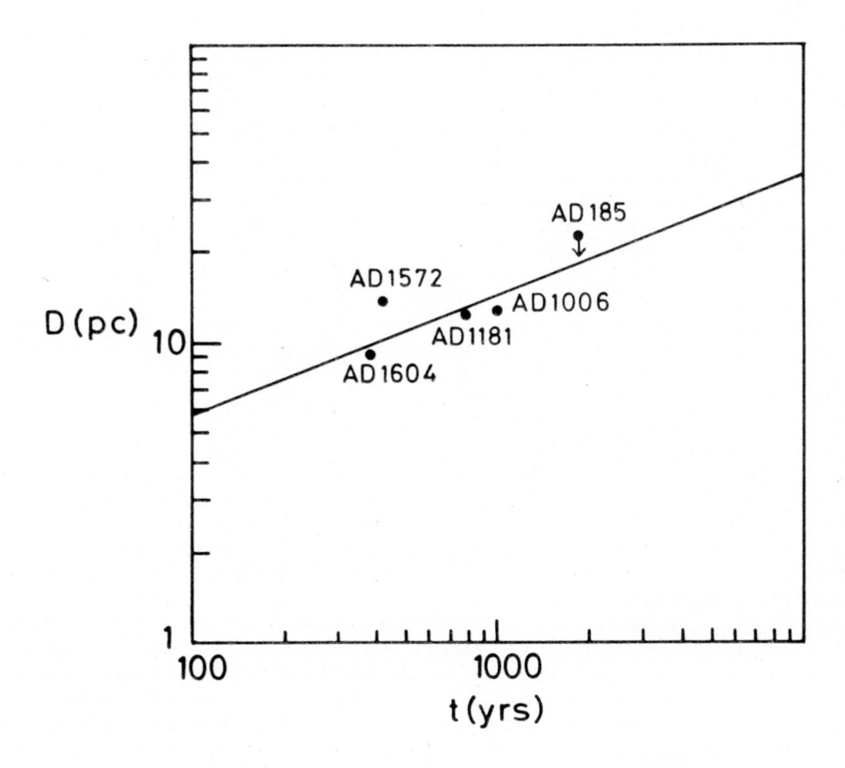

Fig. 12.2. The D-t relationship

evidence of deceleration, so that if this also represents the initial ejection velocity it is almost an order of magnitude less than for other supernovae. The remnant as presently observed at all wavelengths, with central brightening, clearly owes its existence to the continued injection of relativistic particles from its central pulsar, and as outlined in Chapter 8 many of the Crab Nebula's apparently unique properties may be directly attributed to the pulsar. However it is of interest to consider how the expanding shock wave produced by the supernova outburst may have evolved, and whether its effects might be observable.

The results discussed earlier suggest that E_o/n typically is $\sim 5 \times 10^{51}$ erg cm^3. Adopting this value for the supernova of AD 1054 gives $D \sim 14$pc from the Sedov relationship, so that for the usually accepted distance to the Crab Nebula of 2.0 kpc, the expected angular diameter of the shock wave would be ~ 24 arc min., with large uncertainty. (Note that if the outer boundary of the observed radio remnant is regarded as the present diameter of the shock wave, the observed value of E_o/n is only 2×10^{48} erg cm^3. If one was to then assume a value of $n \sim 0.2$ cm^{-3}, the estimated initial blast energy $E_o \sim 10^{47}$ ergs. This is several orders of magnitude lower than the value estimated for any other galactic supernova, and indeed less than the present kinetic energy of the expanding nebula).

The position of the shock wave delineates a shell-like region of radio emission in the usually accepted supernova remnant models. Hazard and Sutton (1971) suggested that the presence of a shell source of radio emission surrounding the prominent central source of the Crab Nebula cannot be excluded since it would be very difficult to detect if its surface brightness were low.

Soft X-ray extensions to the Crab Nebula have been reported by Toor et al (1975) from lunar occultation data, and by Charles and Culhane (1976) from observations with the soft X-ray detector on the satellite Copernicus. We suggest that these extensions may be due to the shock from the supernova of AD 1054. The interstellar material heated by the shock wave expanding from the site of a supernova is expected to be a source of X-rays. Applying the standard adiabatic shock wave model as summarized by Clark and Culhane (1976) to the present shock angular diameter of ~ 24 arc min inferred above for the remnant of AD 1054, and assuming $n \sim 0.2$ cm^{-3} at the z-distance of the Crab Nebula, gives an estimated intrinsic X-ray luminosity of $\sim 10^{34}$ erg s^{-1} and an expected observed flux of $\sim 0.15 \times 10^{-10}$ erg cm^{-2} s^{-1} in the energy interval 0.5 to 1.5 keV. Such weak emission would not have been detectable above background with the 2.1 arc min equivalent beamwidth of Copernicus. We infer therefore that if the soft X-ray extensions actually detected are evidence of the original shock, they must result from interaction of the shock with local interstellar density enhancements - this is the explanation usually invoked for the "patchiness" of emission for all the X-ray supernova remnants. The X-ray extensions were observed by the Copernicus satellite to a diameter of approximately 8 arc min - considerably less than the 24 arc min diameter inferred from the calculations above. However the Copernicus observations did not go beyond 5 arc min of the centre of the source - soft X-ray observations out to 15 arc min may reveal the total extent of the shock wave.

By suggesting that the present position of the shock wave from the supernova of AD 1054 lies beyond the region of optical filamentary structure and beyond the outer boundary of the radio source, many of the apparently anomalous properties of the supernova may be accounted for. Certainly the initial blast energy could then approach the characteristic value of a few 10^{50} ergs. There remains the problem of the low proper motion measured for the optical filaments of the Crab Nebula. The amount of material producing the optical emission lines through ionisation by the synchrotron radiation is thought to be at most about one solar mass (Minkowski, 1968), On the above interpretation for the supernova of AD 1054 this material is <u>not</u> the initial high-velocity ejecta of the supernova explosion. (Ejections of different velocities are recognised as a feature of Type II supernovae). The present kinetic energy of expansion of the nebula of $\sim 10^{49}$ ergs may have been derived at least in part from the loss of radial energy of the pulsar.

It appears that there are a number of remnants within the galaxy that show a general resemblance to the supernova of AD 1054 (Lockhart et al, 1976). It is also possible that the radio source MSH 15 - 56 is a similar object at a much later stage of evolution (see Fig. 12.3). Certainly the radio source shows an amorphous "central" structure like the Crab Nebula; this displays significant polarization indicative of the synchrotron emission process invoked for supernova remnants, and also has an unusually shallow radio spectrum. In the case of MSH 15 - 56 however, the present position of the shock front is clearly delineated by significant peripheral radio emission lying well beyond the amorphous structure (Clark et al, 1975).

Apart from the Crab Nebula, the six historical supernovae individually suggest values of E_o/n quite close to the mean value of 5×10^{51} erg cm^3 described in Clark and Caswell (1976) for SNRs above a given surface brightness threshold. Tycho's SN does however indicate a somewhat above average value for E_o/n. The comparisons are of course fairly crude because they depend heavily on accurate measurements of distance d ($E_o/n \propto d^5$) so that a 15% error in d results in a factor of 2 error in E_o/n.

In the case of the historical supernovae, a fairly small scatter in the observed values of E_o/n cannot be due to a systematic selection effect; furthermore, unless E_o is proportional to n (which seems unlikely) it implies that the range of values of both E_o and n show quite low dispersions. It has previously been suggested that n would show quite strong dependence on z, the perpendicular height above the plane. For example, Clark and Culhane (1976) assumed a horizontally stratified medium with

$$n(z) = n(0) \exp - \frac{z}{z_o} \qquad (8)$$

taking a density on the plane n(0) = 0.5, and a scale height of ~ 120 pc. This dependence on z may have been over estimated

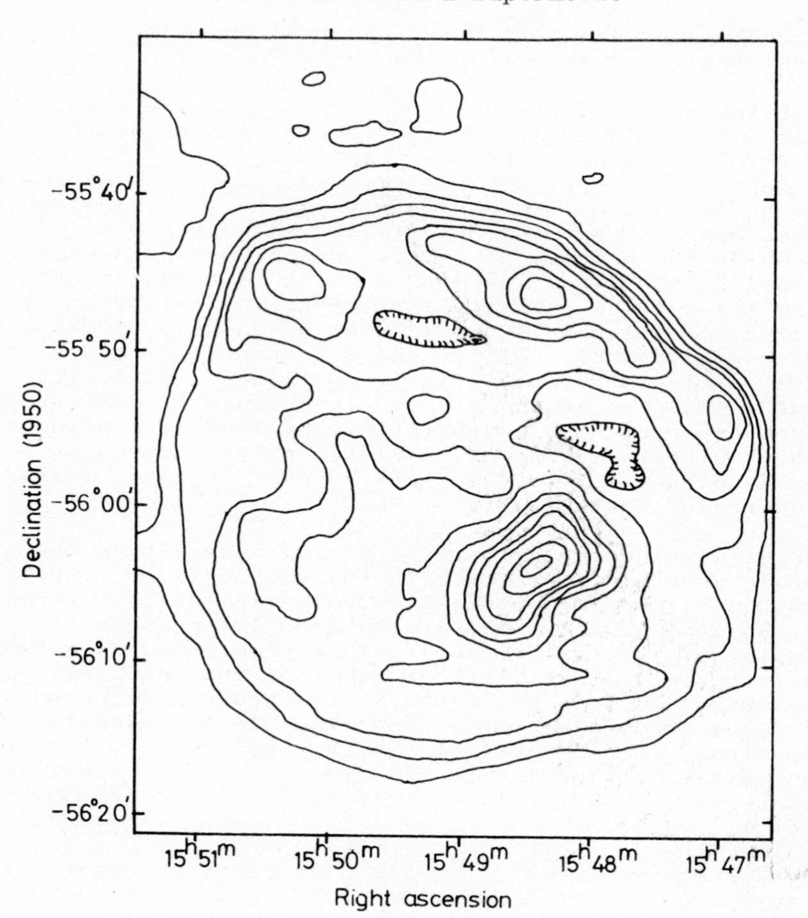

Fig. 12.3. The SNR MSH 15-56.

Consider an interstellar medium comprising both a diffuse gas of density n_{is} and cloudlets of greatly enhanced density . A value of n_{is} (0) as small as 0.2 cm^{-3} may be applicable, and the scale height of the diffuse component is probably much greater than that of the cloud/cloudlet component. We would suggest that the diffuse component should be modelled by equation (8) with a scale height of, say, $z_{is} \sim 1000$pc, with the clouds having mean density given by

$$n_{cl}(z) = n_{cl}(0)\exp - \frac{z}{z_{cl}}$$

where z_{cl} is the scale height ($\ll z_{is}$) of the clouds. z_{cl} might be expected to be ~ 50 pc. If $n_{cl}(0)$ refers to the mean cloud density averaged over the whole plane, then the variation of $n_{cl}(z)$ with z-distance could reflect either the spatial variation of clouds of constant density, or the varying density of individual clouds occupying a constant fraction of space - or a combination of these.

Once the SNR is significantly larger than a typical cloud the shock wave should propagate through the diffuse medium without being affected significantly by the clouds (McKee and Cowie, 1975). Indeed Sgro (1975) has shown that a plane shock will completely recover its form within 2 to 3 cloud diameters of overtaking a cloud. In the Sedov equation (1) n_{is} should thus be used, and the dynamical evolution of remnants will then show little dependence on z up to several hundred pc - within which range most SNRs are found. Thus for $n_{is} \sim 0.2$ the corresponding typical value of E_o would be 10^{51} ergs. From general considerations it seems reasonable that for such a catastrophic event as a supernova, the threshold might be well defined, with evolution beyond that threshold leading inevitably to supernova activity showing quite small dispersions in the explosion parameters.

It has been argued that the radio emission from an SNR is dominated by quite distinct physical processes at various stages of its evolution. According to the model of van de Laan (1962), the interstellar magnetic field compressed by the shock front and cosmic rays may produce radio emission, but this mechanism is only satisfactory after the remnant has passed from its adiabatic phase to a later radiative phase when large compressions are possible. However since at an adiabatic shock front compression by only a factor of 4 is obtained, an alternative means of producing the magnetic field is required for the bright radio emission from young and middle-aged remnants. According to Gull (1973a), Rayleigh-Taylor instabilities at the interface between the ejecta and interstellar medium in a young remnant will result in the formation of a zone of unstable convective mixing. In this zone random motions will tangle any pre-existing "frozen-in" magnetic field producing "knots" of greatly enhanced magnetic field sufficient to describe the strong radio emission from such young SNRs as Cas A and the remnant of Tycho's SN. After a time this convection zone expands and dissipates, so that no middle-aged remnants (age $>$1000 years) would be expected to be radiating by this mechanism alone. Thus in a uniform interstellar medium there is no obvious reason why "middle-aged" SNRs in the adiabatic phase of their evolution should be strong radio emitters. Nevertheless, the radio brightness of SNRs appears to decrease monotonically with time throughout the adiabatic phase, (at least down to a limiting value of surface brightness - see Fig. 12.1).

McKee and Cowie (1975) suggest that the optical emission of supernovae in the adiabatic expansion phase comes principally from interaction of the shock wave with the denser cloudlets of the interstellar medium, and this is also the mechanism usually invoked for the patchiness of X-ray emission. Clark and Caswell (1976) speculated that much of the radio emission may also arise from this interaction. After the shock has crossed a cloud, a Rayleigh-Taylor instability will occur (McKee and Cowie, 1975). This leads to the possibility that in this region for SNRs in the adiabatic phase non-thermal radio emission may occur due to the turbulent amplification of the magnetic field and acceleration of relativistic particles by convective motions, in a similar manner to the mechanism suggested by Gull (1973a) for young SNRs. If, as suggested above, the scale height of dense cloudlets, z_{cl}, is significantly smaller than that of the diffuse medium then SNRs at large values of z may be subluminous in general; however their rate of expansion, dependent on n_{is}, may not differ significantly from that of SNRs close to the galactic plane. (Additionally, SNRs with diameters approaching the scale-height of cloudlets would be expected to be significantly brighter on their edge nearest the plane - as in fact is observed for a significant number of such SNRs. For example, the SNRs shown in Figs. 4.6, 6.4, and 12.3 are all brightest on their edges nearest the plane.)

The above explanation might account for the particularly subluminous remnants of AD 1006 and AD 1054 (in the latter case ignoring the emission still being excited by the central pulsar), both at several cloud scale heights from the plane. Note that such subluminous remnants would be under-represented in a sample selected from radio remnant detections relative to an 'historical' sample (which will in fact favour high latitude objects suffering less obscuration). It seems clear, however, that at least some SNRs at quite large values of z are not subluminous in their radio emission, and in such cases perhaps they occur in regions with more cloudlets than is common at such z values; indeed there may be a distinct bias whereby potential supernovae preferentially occur in such regions with quite high concentrations of dense clouds. Additionally, the longevity of certain SNRs at large z may result from the presence of an active pulsar (Clark, 1976).

We finally consider the apparent turnover in the Σ - D relationship noted in Fig. 12.1. Caswell and Clark (1976) slightly revised the best estimate of this change in slope, proposing that $\Sigma_{408} = 10^{-15}$ D^{-3} be continued to D = 38pc rather than 32 pc; at 38pc $\Sigma \sim 1.8 \times 10^{-20}$ Wm^{-2} Hz^{-1}, the age will be $\sim$10,700 years, and the expansion velocity still $\sim$1000 km s^{-1}. The Σ - D slope then appears to steepen (to a slope of $\sim -$ 10) and since the expansion velocity is still too high to interpret the slope change as a transition to the isothermal phase, Caswell and Clark suggested that it is due to a rapid decline in the efficiency of the radio emission mechanism suggested above when the shock velocity falls below about 1000 km s^{-1}. In view of the rather speculative nature of the proposed emission mechanism, this suggestion is not an obvious conclusion but rather, it is hoped, another clue to understanding the factors determining the radio intensity.

On this interpretation, the adiabatic expansion of the shock continues beyond the diameter at which the radio SNRs essentially fade below the limit of detectability, i.e. to $D \sim 65$ pc. Thus if D can be estimated by any means (including the use of the Σ - D curve after its turnover to the steep slope of -10), then application of the D-t relationship will probably give a reasonably good estimate of age.

From the above discussion, it seems likely that most of the radio SNRs may be regarded as similar objects following a common dynamical evolutionary track. Dense cloudlets in the interstellar medium may play an important role in the radio brightness of a particular remnant, although variations in the environment (of a magnitude commonly encountered) have only slight effects on the dynamical evolution.

At the beginning of this book we included a quotation from Zwicky (1965): "The investigation of the remnants of supernovae and their relation to historical records, both written and unwritten, will be one of the most fascinating tasks awaiting the next generation of astronomers." The quotation goes on - "These records may also include various features of a geological and a biological nature, refering back to evolutionary phenomena of the Earth's surface and the life on it. Already a book could be written about the subject of the remnants of supernovae and about the efforts which have been made to elucidate their characteristics.".

In writing this present book we have limited ourselves to the written historical records of supernovae, simply because this is our own particular field of expertise, experience and interest. We have certainly found, as Zwicky predicted, the detailed investigation of the written records "fascinating". The undoubtedly equally fascinating task of investigating the unwritten records must await the attention of others. All we will do here is to speculate on the possible scope of such investigations, and the possible effects of a "nearby" supernova on the Earth and its environment.

McCrea (1975a, b) has proposed that the encounter of the solar system with a dense cloud of interstellar material during its passage through a spiral arm of the Galaxy may produce such climatic catastrophes on Earth as the ice epochs. His thesis, an extension of several earlier investigations is based on the expected affect on the solar constant of an increased accretion rate. Unfortunately the cloud density necessary to produce the required variation is 10^5 - 10^7 cm^{-3}, and although clouds with such extreme densities are thought to exist, Begelman and Rees (1976) pointed out that they are so compact and unusual as to reduce the likelihood of the Sun ever encountering one. Begelman and Rees in fact showed that a much more modest cloud density (10^2 - 10^3 cm^{-3}) would prevent the solar wind from reaching the Earth, with resulting modification of the near-Earth environment; however the climatic consequences of such a modification are not well understood. We present here an alternative hypothesis which still retains McCrea's association with the passage of the solar system through a spiral arm, but which relates the initiation of an ice epoch, plus biological catastrophes, to an encounter of the solar system with a nearby supernova outburst rather than a superdense interstellar cloud.

The spiral structure of our galaxy indicates the regions where bright, short-lived stars illuminate the inter-stellar gas and dust. The two principal spiral arms are believed to reflect a self-sustaining density-wave pattern through which the disk of stars and gas move. Observations indicate dust lanes along the inner (concave) edge of each arm, and this is interpreted as a shock-wave effect whereby clouds become temporarily compressed as they cross the lane. A cloud of interstellar matter in its orbit round the Galaxy is liable to enter a spiral arm via its compression lane about once in $\sim 10^8$ y. In general the cloud material will re-expand on emerging from the lane after $\sim 10^6$ y (the cloud not necessarily retaining its original identity). But sometimes a cloud is compressed to a condition for star formation, and the result is a new galactic cluster which proceeds to traverse the associated spiral arm itself, taking $\sim 10^7$ y to do so. If this cluster contains a star of mass some 10 $M_\odot$ or greater it would be expected to evolve rapidly over less than $\sim 10^7$ y, terminating in a supernova outburst of Type II before emerging from the spiral arm in which it is formed. In this view supernovae of Type II would occur only in spiral arms (or near galactic nuclei), in accord with observation. Indeed a "supernova zone" might be defined lying beyond the compression lane of a spiral arm; such a zone is depicted diagrammatically in Fig. 12.4, on the assumption of a 20 $M_\odot$ stellar life-time of $\sim 5 \times 10^6$ y and a transverse velocity (at about the solar distance from the galactic centre) of 200 km/s giving a zone width of ~ 1 kpc.

Clark and Caswell (1976) have concluded from a statistical analysis of the radio remnants of supernovae that the average interval between outbursts leaving long-lived observable remnants (i.e. those lying within spiral arms and evolving only slowly) is 100 years. Shklovsky (1968) prefers an interval of a mere 30 years for Type I (Population 2) events, but accepts a longer interval of ~ 100 years for those of Type II (Population 1). Results from other investigators range between these extremes. In the absence of a more definite estimate, we will take an average interval between Type II supernovae of 100 y.

If there is one Type II supernova per 100 y, and if, as seems reasonable, this is localized to one of the "supernova zones" proposed above, then this implies 5×10^4 such supernovae in one (of two) spiral arms in a period of $\sim 10^7$ y (the time for the Sun to traverse a spiral arm). Since a spiral arm is $\sim 2 \times 10^4$ pc long, this suggests that 50 supernovae will occur in a 20 pc strip of a spiral arm's "supernova zone" each time the Sun crosses it. Thus <u>at least one</u> supernova would be expected to occur within 10 pc of the Sun <u>each time</u> it crosses a spiral arm (i.e about once every 10^8 years). Note that this simple calculation takes no account of the thickness of a spiral arm perpendicular to the plane; however since the bulk of supernova remnants are concentrated to within a few tens of pc of the plane, this simplification is not expected to increase the above estimate for the interval between nearby supernovae by more than a factor of two.

A similar estimate is obtained from an interpretation of the galactic distribution of supernova remnants. Most supernovae appear to be restricted to within little more than a solar distance of the galactic centre and close to the plane i.e. to a volume of $\sim 10^{10}$ cubic pc. Making the simplifying assumption that

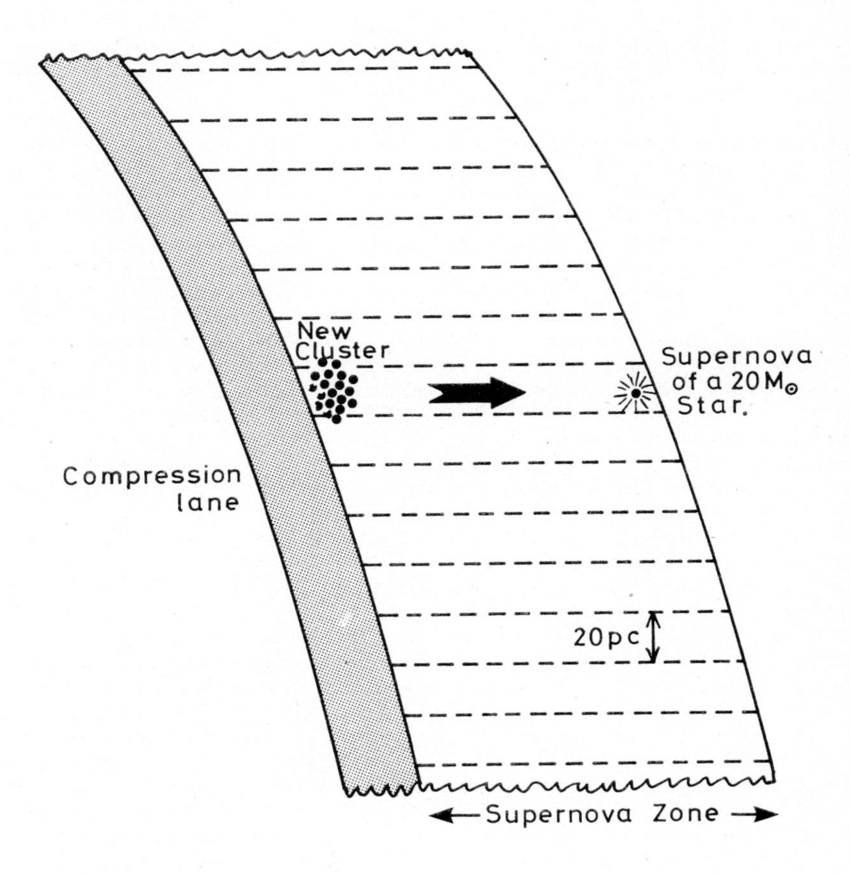

Fig. 12.4. A schematic representation of part of a spiral arm, showing a "supernova zone".

the rate of supernovae per unit volume is uniform throughout the Galaxy, then the interval between nearby supernovae occurring within distance R of the Sun is

$$T = \frac{\tau \, 10^{10}}{4/3 \, \pi \, R^3} \quad y$$

where τ y is the characteristic interval between supernovae. With R $\sim$10 pc, T $\sim$2.5 x 10^8 y - of the same order as the value obtained above. One must thus conclude that during the history of our solar system ($\sim$5 x 10^9 y) more than 20 supernovae would be expected to have occurred within a distance of 10 pc, and over 2500 within a distance of 50 pc of the Sun.

Estimates of the radiation flash (see for example Colgate, 1968) emitted (mainly in γ and X-rays) in a supernova explosion vary from $\sim10^{47}$ to 10^{50} ergs. (This represents a very small fraction of the total energy of the outburst of $\sim10^{52}$ ergs). Assuming the lower value, the peak integrated energy flux expected above the Earth's atmosphere from a supernova at $\sim$10 pc distance is 10^6 erg cm^{-2} - this compares with the solar flux of $\sim$20 erg/cm^2 sec in the wavelength range of interest. Absorption in the upper atmosphere means that this flash gives no significant surface radiation dose initially. Nevertheless Ruderman (1974) has predicted catastrophic effects on the Earth's protective ozone layer. The ionizing radiation from the supernova flash produces excited free nitrogen atoms in the atmosphere which quickly oxidize. The oxides of nitrogen catalytically destroy ozone, and Ruderman claims that by this mechanism the ozone layer density may be depleted to just a few percent for the order of a hundred years. Such destruction of the ozone layer would expose the Earth's surface, and more particularly life on it during the past 600 million years, to lethal solar ultraviolet radiation.

The expanding shell of ejecta from a supernova at 10 pc distance would reach the Earth after about 2000 years, with the passage of the shell taking several hundred years. During this latter period the density of primary cosmic rays would be expected to be enhanced by perhaps a factor of a hundred (Shklovsky, 1968). About one third of the background radio-activity level at the Earth's surface is contributed by cosmic radiation (the remainder being due to terrestrial factors). Thus a hundredfold increase in cosmic ray intensity would produce a drastic thirty-fold increase in mean level of radioactivity. In addition, the enhanced cosmic ray flux would be expected to produce the same destruction of the ozone layer as produced by the initial radiation flash. The possible biological consequences of such dramatic increases in background radioactivity and ultraviolet radiation fluxes expected for protracted periods sometime after a nearby supernova outburst may have been catastrophic for certain species. Ruderman has suggested that in man, the hugh increase in ultraviolet radiation would give rise to a vast increase in the incidence of skin cancer; in animals with calcified internal skeletons the increased vitamin D production might have been toxic.

The supernova-initiated destruction of the ozone layer would be expected to have climatic as well as biological consequences. The long-term reduction of ozone density to possibly just a few percent of its normal value would lead to an increase of several percent in solar energy transmitted to the ground, initially producing an increase of several degrees in mean temperature. (It is interesting to note how many of the historical records of the AD 1006 supernova, the closest to the solar system in historical times, refer to it being followed by drought, famine, and pestilence. However at a distance of $\sim$1 kpc the expected energy flux at the Earth would be only one ten thousanth that of a supernova at 10 pc.) The subsequent increased precipitation and cloud cover may eventually result in the reflection of an increased proportion of incident energy and a lowering of mean temperature at the Earth's surface. Could nearby (closer than 10 pc) supernovae, occurring on average every few hundred million years, have initiated the climatic conditions leading to the onset of the ice-ages on Earth? We believe that this is a possibility worthy of further consideration.

Possible geological records of nearby supernovae are difficult to envisage. The increased primary cosmic ray flux during the passage of a supernova remnant shell may have sufficiently modified the Earth's radiation belts and consequently the geomagnetic field to be revealed in palaeomagnetic data. Energetic cosmic ray protons might be expected to produce proton-rich isotopes by the so-called (p,n) reaction (the ejection of a neutron from an atomic nucleus following bombardment by a proton), which may show up for certain epochs in core samples from polar regions; (at middle and low latitudes the Earth's atmosphere and magnetic field absorb or deflect nearly all primary cosmic rays.) A more likely site where such records might be preserved would be a still-active planet with negligible atmosphere or magnetic field, and Mars is the obvious example.

The final, and what must be the ultimate hypothesis relating to the consequences of supernovae must be left to Shklovsky (1968): "We can imagine that a high level of radioactivity, caused by some cosmic events which took place in an epoch some 10^9 years ago, may have stimulated the formation of highly complex compounds from simple organic compounds, and life on the Earth may have developed on the basis of these complexes."

Perhaps we have wandered too far into the realms of speculation. We will conclude our present study with a final consideration of the ancient written records of supernovae.

Astronomy has reaped enormous benefit from the literary traditions of China, and its richly recorded history. The Far-Eastern astronomical records represent a unique collection of data from a full two millenia, and the essential reliability of the observations (despite political, social, and astrological pressures) are beyond dispute. Duplication of certain observations from the Arab lands after about the 9th century AD provide useful confirmatory results. The post-Renaissance contributions from Europe may have surpassed the ancient observations in precision , but not in their impact on modern astrophysical investigations. We can only reiterate and reinforce the claim we made at the end of Chapter 1, that there can be little doubt that the historical astronomical records must be regarded as among the most valuable legacies which the ancient world has bequeathed to modern science.

REFERENCES

The following abbreviations are used for frequently occurring reference sources :-

A.A.	Astronomy and Astrophysics
A.J.	Astronomical Journal
A.J.P.	Australian Journal of Physics
A.N.	Astronomische Nächrichten
A.P.A.E.	Astronomical Papers of the American Ephemeris
Ap. J.	Astronomical Journal
Ap. Letters	Astrophysical Letters
M.R.A.S.	Memoirs of the Royal Astronomical Society
M.N.R.A.S.	Monthly Notices of the Royal Astronomical Society
P.A.S.P.	Publications of the Astronomical Society of the Pacific
Q.J.R.A.S.	Quarterly Journal of the Royal Astronomical Society

Allen, C.W., 1973. Astrophysical Quantities (3rd Edition) (Univ. of London Press).

Argelander, F., 1864. A.N., 62, 273.

Arnold, T., 1885. Simeonis monachi Dunelmensis Historia Regum (HMSO, London).

Baade, W., 1942. Ap. J., 96, 188.

Baade, W., 1943. Ap. J., 97, 119.

Baade, W., 1945. Ap. J., 102, 309.

Baldwin, J.E., 1971. In The Crab Nebula (Reidel, Holland).

Barbon, R., Ciatti, F., and Rosino, L., 1974 a and b. In Supernova and Supernova Remnants (Reidel, Holland).

Bemporad, A., 1904. Mitt. Heidelberg, No. 4.

Bielenstein, H., 1950. Bull. Mus. Far. E. Antiq., 22, 127.

Biot, E., 1843. Connaissance des Temps., Additions, p60.

Böhme, S., 1936. A.N., 262, 479.

Bolton, J.G., 1948. Nature, 162, 141.

Bolton, J.G., and Stanley, G.J., 1949. Australian. J. Sci. Res., 2A, 139.

Bowyer, C.S., et al., 1964. Science, 146, 912.

Brandt, J.C., et al., 1975. In Archaeoastronomy in Pre-Columbian America. ed. A.F. Aveni. (Univ. of Texas Press).

Brosche, P., 1967. In Variable Stars. IAU Info. Bull. No. 192.

Caswell, J.L., 1967. M.N.R.A.S., 136, 11.
Caswell, J.L., 1970. A.J.P., 23, 105.
Caswell, J.L., and Clark, D.H., 1975. A.J.P. Astrophys. Suppl., No. 37, 57.
Caswell, J.L., Clark, D.H., and Crawford, D.F., 1975. A.J.P. Astrophys. Suppl., No. 37, 39.
Caswell, J.L., et al., 1976. A.A. - in press.
Caswell, J.L., and Clark, D.H., 1976. M.N.R.A.S. - in press.
Chambers, G.F., 1909. The Story of Comets. (Clarendon Press, Oxford.)
Charles, P.A., and Culhane, J.L., 1975. Scientific American, 233, 38.
Ch'en-yuan, 1935. Sung-hui-yao-chi-kao (Peking).
Chevalier, R.A., 1974. Ap. J., 188, 501.
Chu Sun-il, 1968. J. Korean Astr. Soc., 1, 29.
Clark, D.H., 1976. M.N.R.A.S., 175, 77p.
Clark, D.H., Caswell, J.L., and Green, A., 1973. Nature, 246, 28.
Clark, D.H., Caswell, J.L., and Green, A., 1975. A.J.P. Astrophys. Suppl., No. 37, 1.
Clark, D.H., Green, A., and Caswell, J.L., 1975. A.J.P. Astrophys. Suppl. No. 37, 75.
Clark, D.H., and Caswell, J.L., 1976. M.N.R.A.S., 174, 267.
Clark, D.H., and Culhane, J.L., 1976. M.N.R.A.S., 175, 573.
Clark, D.H., and Stephenson, F.R., 1976. Q.J.R.A.S., 17, 290.
Davison, K., and Tucker, W., 1970. Ap. J., 161, 437.
Davison, P.J.N., Culhane, J.L., and Mitchell, R.J., 1976. Ap. J., 206, L37.
Day, G.A., Caswell, J.L., and Cooke, D.J., 1972. A.J.P. Astrophys. Suppl., No. 25, 1.
Dickel, J.R., McGuire, J.P., and Yang, K.S., 1965. Ap. J., 142, 798.
Dickel, J.R., and Yang, K.S., 1965. Ap. J., 142, 1642.
Dickel, J.R., and McKinley, R.R., 1969. Ap. J., 155, 67.
Dickel, J.R., Milne, D.K., Kerr, A.R., and Ables, J.G., 1973. A.J.P., 26, 379.
Dicks, D.R., 1960. The Geographical Fragments of Hipparchus. (Univ. of London Press.)
Dombrovsky, V.A., 1954. Akad. Nauk. USSR., 94, 1021.
Downes, D., 1971. A.J., 76, 305.
Dreyer, J.L.E., 1890. Tycho Brahe (Adam and Charles Black, Edinburgh).
Dreyer, J.L.E., 1906. History of the Planetary System from Thales to Kepler. (Cambridge Univ. Press).
Duin, R.M., and Strom, R.G., 1975. A.A., 39, 33.
Duin, R.M., et al., 1975. A.A., 38, 461.
Duncan, J.C., 1921. Proc. Nat. Acad. Sci., 7, 179.
Duncan, J.C., 1939. Ap. J., 89, 482.
Duyvendak, J.J.L., 1942. P.A.S.P., 54, 91.
Ellis, F.H., 1975, In Archaeoastronomy in Pre-Columbian America. ed. A.F. Aveni. (Univ. of Texas Press).
Ewen, H.I., and Purcell, E.M., 1951. Nature, 168, 356.
Fessenkov, V.G., 1962. In Physics and Astronomy of the Moon. ed. Z. Kopal. (Academic Press, New York).
Fotheringham, J.K., 1919. M.N.R.A.S., 79, 162.

Friedman, H., Byram, E.T., and Chubb, T.A., 1967. Science, 156, 374.
Goldstein, B.R., 1965. A.J., 70, 105.
Goldstein, B.R., 1966. Nature, 211, 504.
Goldstein, B.R., and Ho Peng Yoke, 1965. A.J., 70, 748.
Goss, W.M., and Schwarz, U.J., 1971. Nature, 234, 52.
Goss, W.M., et al., 1972. Ap. J. Suppl., 24, 123.
Goss, W.M., Schwarz, U.J., and Wesselius, P.R., 1973. A.A., 28, 305.
Green, A.J., 1971. A.J.P., 24, 773.
Green, A.J., 1974. A.A. Suppl., 18, 325.
Gull, S.F., 1973a. M.N.R.A.S., 161, 47.
Gull, S.F., 1973b. M.N.R.A.S., 162, 135.
Gull, S.F., 1975. M.N.R.A.S., 171, 237.
Hagen, J.P., and McClain, E.F., 1954. Ap. J., 120, 368.
Han Yu-shan, 1955. Elements of Chinese Historiography. (W.M. Hawley, Hollywood).
Hanbury-Brown, R., and Hazard, C., 1952. Nature, 170, 364.
Haslam, C.G.T., and Salter, C.J., 1971. M.N.R.A.S., 151, 385.
Hawkins, F.J., et al., 1974. M.N.R.A.S., 169, 41.
Hazard, C., and Sutton, J., 1971. Ap. Letters, 7, 179.
Hellman, C.D., 1960. Isis, 51, 322.
Henning, K., and Wendker, H.J., 1975. A.A., 44, 91.
Herman, B.R., and Dickel, J.R., 1973. A.J., 78, 879.
Hewish, A., and Okoye, S.E., 1964. Nature, 203, 171.
Higgs, L.A., and Halperin, W., 1968. M.N.R.A.S., 141, 209.
Hill, E.R., 1964. Symp. IAU-URSI, No. 20., 107.
Hill, E.R., 1967. A.J.P., 20, 297.
Hill, I.E., 1972. M.N.R.A.S., 157, 419.
Ho Peng Yoke, 1962. Vistas in Astronomy, 5, 127.
Ho Peng Yoke. 1966. The Astronomical Chapters of the Chin Shu. (Mouton, Paris.)
Ho Peng Yoke, 1969. J. Asian History, 3, 135.
Ho Peng Yoke, 1970. Oriens Extremus, 17, 63.
Ho Peng Yoke, Paar, T.H., and Parson, P.W., 1970. Vistas in Astronomy, 13, 1.
Hsi Tsê-tsung, 1955. Acta Astr. Sinica, 3, 183. Trans. 1958 in Smithson. Contr. Astrophys., 2, 109.
Hsi Tsê-tsung, and Po Shu-jen, 1965. Acta. Astr. Sinica, 13, 1. Trans. NASA Tech. Trans. TTF-388. Abridged trans. by K.S. Yang, 1966 in Science, 154, 597.
Hsüeh Chung-san and Ou-yang, I, 1956. A Sino-Western Calendar for Two Thousand Years, AD 1-2000, (Peking).
Hughes, M.P., Thompson, A.P., and Coluin, R.S., 1971. Ap. J. Suppl., 23, 323.
Humboldt, A. von., 1851. Cosmos. (English trans. by E.C. Otte), vol. III, (Bohn, London).
Ilovaisky, S.A., and Lequeux, J., 1972. A.A., 18, 169.
Isles, J.E., 1976. J. Brit. Astr. Assn., 86, 245.
Jao Tsung-i, 1959. Yin-tai-cheng-pu-jen-wu-t'ung-kao (Hong Kong Univ. Press).
Jenkins, L.F., 1952. General Catalogue of Trigonometric Stellar Parallaxes. (Yale Univ. Observatory, New Haven).
Jones, J., Spencer, 1961. General Astronomy, (4th Edition), (Edward Arnold Ltd., London).

Jones, B.B., 1973. A.J.P., 26, 545.
Kanda Shigeru, 1935. Nihon Temmon Shiryo. (Koseisha, Tokyo)
Keen, N.J., et al., 1973. A.A., 28, 197.
Kerr, A.R., 1968. In Nebulae and Interstellar Matter, Vol. VII, of Stars and Stellar Systems. (Chicago).
Kestenbaum, H., Angel, J.P.P., and Novick, R., 1971. Ap. J., 164, L87.
Kiang, T., 1972. M.R.A.S., 76, 27.
Kin Yong-woon, 1974. Korea Journal, 14, 4.
Kukarkin, B.V., et al., 1969. General Catalogue of Variable Stars. (Moscow).
Kundu, M.R., 1971. Ap. J., 165, L55.
Kundu, M.R., and Velusamy, T., 1972. A.A., 20, 237.
Lundmark, K., 1921. P.A.S.P., 33, 225.
Lynn, W.T., 1884. Observatory, 7, 17 and 75.
Mayall, N.U., 1937. P.A.S.P., 49, 104.
Mayall, N.U., and Oort, J.H., 1942. P.A.S.P., 54, 95.
Mayer, C.H.T., McCullough, T.P., and Sloanaker, R.M., 1957. Ap. J., 126, 468.
McKee, C.F., 1974. Ap. J., 188, 335.M
McKee, C.F., and Cowie, L.L., 1975. Ap. J., 195, 715.
Menon, T.K., and Williams, D.R.W., 1966. A.J., 71, 392.
Miller, W.C., 1955a. Plateau, Mus. of N. Arizona, 27, 6.
Miller, W.C., 1955b. Astron. Soc. Pacific, Leaflet No, 314.
Mills, B.Y., Slee, O.B., and Hill, E.R., 1957. A.J.P., 11, 360.
Mills, B.Y., 1974. In Galactic Radio Astronomy. (Reidel, Holland)
Milne, D.K., 1969. A.J.P., 22, 613.
Milne, D.K., 1970. A.J.P., 23, 425.
Milne, D.K., 1971a. A.J.P., 24, 429.
Milne, D.K., 1971b. A.J.P., 24, 757
Milne, D.K., 1972. Ap. Letters, 11, 167.
Milne, D.K., and Hill, E.R., 1969. A.J.P., 22, 211.
Milne, D.K., and Dickel, J.R., 1975. A.J.P., 28, 209.
Minkowski, R., 1942. Ap. J., 96, 199.
Minkowski, R., 1943. Ap. J., 97, 128.
Minkowski, R., 1959. Paris Symposium on Radio Astronomy. (Stanford Univ. Press).
Minkowski, R., 1964. Ann. Rev. Astr. Ap., 2, 247.
Minkowski, R., 1966. A.J., 71, 371.
Minkowski, R., 1968. In Nebulae and Interstellar Matter, Vol. VII. of Stars and Stellar Systems. (Chicago).
Minkowski, R., 1971. In The Crab Nebula. (Reidel, Holland).
Morohashi Tetsuji, 1955. Dai Kanwa Jiten (Tokyo).
Moule, A.C., and Yetts, W.P., 1957. The Rulers of China. (Routledge and Kegan Paul, London).
Muller, P.M., and Stephenson, F.R., 1975. In Growth Rhythms and the History of the Earth's Rotation. (John Wiley and Sons Ltd., London).
Muratori, L.A. (ed) 1723. Rerum Italicarum Scriptores, 25 vols. (Milan).
Needham, J., 1959. Science and Civilisation in China, vol. 3. (Cambridge Univ. Press).
Needham, J., Wang-ling, and Price, D. de S., 1960. Heavenly Clockwork, (Cambridge Univ. Press).
Newcomb, S., 1895. A.P.A.E., 6.

Newcomb, S., 1910. A.P.A.E., 9.
Newton, R.R., 1972. Medieval Chronicles and the Rotation of the Earth. (The John Hopkins Univ. Press).
Oppolzer, T.R., von, 1887. Canon der Finsternisse. (Kaiserlichen Akadamie der Wissenschaften, Wien).
Payne - Gaposchkin, Co, 1956. Introduction to Astronomy. (Eyre and Spottiswoode, London).
Payne - Gaposchkin, C., 1957. The Galactic Novae. (North Holland Publishing Co., Amsterdam).
Pertz, G.H., (ed)., 1826. Monumenta Germaniae Historica Scriptores, 32 vols. (Hahn, Hanover).
Pingré, A.G., 1783. Cometographie ou Traite Historique et Theorie des Comets, vol. 1 (Paris).
Porter, N.A., 1974. J. Hist. Astr., 5, 99.
Pskovskii, Y.P., 1963. Sov. Astr., 7, 501.
Pskovskii, Y.P., 1972. Sov. Astr., 16, 23.
Radhakrishnan, V., et al., 1972. Ap. J. Suppl., 24, 49.
Rigge, W.F., 1915. Pop. Astr., 23, 29.
Rosenberg, I., 1970. M.N.R.A.S., 151, 109.
Roger, R.S., Bridle, A.H., Costain, C.H., 1973. A.J., 78, 1030.
Rosenberg, I., and Scheur, P.A.G., 1973. M.N.R.A.S., 161, 27.
Ruderman, M.A., 1974. Science, 184, 1079.
Rufus, W.C., 1913. J. Roy. Asiatic Soc. (Korea Branch), 4, 23.
Rufus, W.C., and Chao, C., 1944. Isis, 35, 316.
Rufus, W.C., and Tien Hsing-chih, 1945. The Soochow Astronomical Chart. (Univ. of Michigan Press, An Arbor.).
Schlier, O., 1934. A.N., 254, 181.
Schlesinger, F., 1935. General Catalogue of Stellar Parallaxes (Yale Univ. Observatory, New Haven).
Schlesinger, F., 1940. Catalogue of Bright Stars, (Yale Univ. Observatory, New Haven).
Schoenberg, E., 1929. In Handbuch der Astrophysik II.
Sedov, L.I., 1959. Similarity and Dimensional Methods in Mechanics, (Academic Press, New York).
Seward, F., et al., 1976. Ap. J., 205, 238.
Sgro., A.G., 1975. Ap. J., 197, 621.
Shower, P.A., 1969. M.N.R.A.S., 142, 273.
Shaver, P.A., and Goss, W.M., 1970. A.J.P., Astrophys. Supple., 14, 77.
Shinoda Minoru, 1960. The Founding of the Kamakura Shogunate 1180 - 1185. (Columbia Univ. Press, New York).
Shklovsky, I.S., 1953. D.A.N., 90, 983.
Shklovsky, I.S., 1960a. Sov. Astr., 4, 243.
Shklovsky, I.S., 1960b. Sov. Astr., 4, 355.
Shklovsky, I.S., 1962. Sov. Astr., 6, 162.
Shklovsky, I.S., 1966. Sov. Astr., 10, 6.
Shklovsky, I.S., 1968. Supernovae, (John Wiley and Sons Ltd., London).
Shklovsky, I.S., 1974. Sov. Astr., 18, 1.
Slee, O.B., and Dulk, G.A., 1974. In Galactic Radioastronomy (Reidel, Holland).
Staelin, D.H., and Reifenstein, E.C., 1968. Science, 162, 1481.
Strom, R.G., and Duin, R.M., 1973. A.A., 25, 351.
Stephenson, F.R., 1971. Q.J.R.A.S., 12, 10.

Stephenson, F.R., 1975. In Origin of Cosmic Rays, (Reidel, Holland).
Stothers, R., 1976. In preparation.
Tammann, G.A., 1966. Nature, 210, 511.
Tchang, M., 1905. Varietes Sinologiques, 24. (Shanghai).
Toor, A., et al., 1976. Ap. J. In press.
Tuckerman, B., 1964. Mem. Amer. Phil. Soc., 59.
Trimble, V., 1968. A.J., 73, 535.
Trimble, V., 1971. In The Crab Nebula, (Reidel, Holland).
Tsui Chi, 1947. A Short History of Chinese Civilisation, (Victor Gollancz, London).
Van den Bergh, S., 1970. Nature, 225, 502.
Van den Bergh, S., Marscher, A.P., and Terzian, Y., 1973. Ap. J.,Suppl., 26, 19.
Van der Laan, S., 1962. M.N.R.A.S., 124, 179.
Velusamy, T., and Kundu, M.R., 1974. A.A., 32, 375.
Vettolani, G., and Zamorani, G., 1976. In preparation.
Weiler, K.W., and Seielstad, G.A., 1971. Ap. J., 163, 455.
Westerlund, B.E., 1969. A.J., 74, 879.
Whiteoak, J.B., and Gardner, F.F., 1968. Ap. J., 154, 807.
Williams, D.R.W., 1973. A.A., 28, 309.
Williams, J., 1871. Chinese Observations of Comets, (London).
Willis, A.G., 1973. A.A., 26, 237.
Wilson, R.A., 1971. In The Crab Nebula (Reidel, Holland).
Wilson, T.L., 1970. Ap. Lett., 1, 95.
Winkler, P.F., and Laird, F.N., 1976. Ap. J. In press.
Woltjer, L., 1964. Ap. J., 146, 1309.
Woltjer, L., 1970. In Interstellar Gas Dynamics, (Reidel, Holland).
Woltjer, L., 1971. In The Crab Nebula, (Reidel, Holland).
Woltjer, L., 1972. Ann. Rev. Astr. Astrophys., 10, 129.
Yabuuchi Kiyoshi, 1967. Sogen Jidai no Kagadu Gijutsu Shi, (Kyoto).
Yang, K.S., 1966. Science, 154, 597.
Zwicky, F., 1939. Phys. Rev., 55, 986.
Zwicky, F., 1965. In Stellar Structure, Vol. VIII of Stars and Stellar Systems, (Chicago)

INDEX